Laboratory Exercises in Oceanography

Harold V. Thurman
Mt. San Antonio College

Samuel M. Savin
Case Western Reserve University

Prentice Hall
Upper Saddle River, New Jersey 07458

Cover Photo: © H. Richard Johnston/Tony Stone Images
Editor: Robert A. McConnin
Production Editor: Mary Harlan
Text Designer: Rebecca Bobb
Cover Designer: Julia Zonneveld Van Hook
Production Buyer: Laura Messerly
Illustrations: Academy Artworks

This book was set in Helvetica by Beth Dubberley

Upper Saddle River, New Jersey 07458

Printed in the United States of America
20 2022

ISBN 0-02-420806-X

Prentice-Hall International (UK) Limited, London
Prentice-Hall of Australia Pty. Limited, Sydney
Prentice-Hall Canada Inc., Toronto
Prentice-Hall Hispanoamericana, S.A., Mexico
Prentice-Hall of India Private Limited, New Delhi
Prentice-Hall of Japan, Inc., Tokyo
Prentice-Hall Asia Pte. Ltd., Singapore
Editoria Prentice-Hall do Brasil, Ltda., Rio De Janeiro

Preface

This laboratory manual was developed to meet the needs of an increasing number of introductory oceanography programs that include a laboratory component. Most of the exercises are also appropriate as homework assignments in courses that include a component of independent problem solving. Used in either manner, the exercises stress both the reinforcement of basic oceanographic principles and terminology and the integration of these principles so that the student may apply them in new situations. The exercises have been written by Samuel M. Savin. They have been developed primarily from homework exercises used in his classes at Case Western Reserve University and from laboratory exercises used by Harold V. Thurman in his course at Mt. San Antonio College.

This preface includes a table that indicates the correspondence between these exercises and appropriate chapters in Thurman's *Introductory Oceanography*, Thurman's *Essentials of Oceanography*, and other college-level introductory textbooks.

The authors are grateful to the large numbers of students who have, over the years, offered many suggestions for improvement to earlier versions of these exercises. Dean A. Dunn, Tom S. Garrison, and Lawrence Krissek provided us with helpful reviews of the manuscript.

A brief overview of each exercise is given below.

Exercise 1, The Earth and the Oceans—Basic Concepts, introduces the student to the metric system, scientific notation, and basic concepts of the earth's shape and size. It includes determination of latitude, determination of the earth's circumference, oceanic dimensions and the mass and density of the earth.

Exercise 2, Maps, Charts, and Marine Geography, introduces longitude determination and map projections. Students gain experience in plotting locations by longitude and latitude, understanding great circle routes, global distribution of land and water, and hypsographic analysis. A drill on the locations of important geographic features is included.

Exercise 3, Bathymetry, Charts, and Navigation, introduces hydrographic charts. It includes exercises that call for measuring distance, determining speed, plotting ship courses, contouring water depth, and constructing a bathymetric cross section.

Exercise 4, Magnetics and Plate Motion Reconstructions, introduces the magnetic field of the earth and paleomagnetics as they relate to seafloor spreading. The student investigates the mirror image characteristics of measurements of the intensity of the earth's magnetic field measured along profiles across spreading centers, estimates the rate of seafloor spreading in the Atlantic using sediment age vs. distance relations and the rate of plate motion in the Pacific using evidence related to the Hawaii-Emperor Seamount hot spot.

Exercise 5, Submarine Topography, Isostasy, Seismicity, and Plates, illustrates the relationship of submarine topography and isostasy to plate tectonics. Exercises investigate the growing and shrinking of ocean basins, abyssal plain development, relationship of spreading rate to steepness of mid-ocean ridge slopes, and the distribution of earthquakes associated with subduction zones.

Exercise 6, Ocean Sediments, examines the relationship between deep-sea geological processes and the nature of ocean sediments. Students may complete the exercise by using a data set provided, or they may obtain their own data by processing sediment samples using a sieve set. Histograms of particle size distribution are constructed in order to compare the degree of sorting of two sediment samples. The velocity of the Grand Banks turbidity current is estimated from data on the timing of the cable breaks that occurred.

Exercise 7, Chemical Aspects of Seawater, examines the origins of and the processes that determine the concentrations of constituents of seawater. Exercises investigate the possible water content of the earth's mantle and whether it could have been the source of ocean water, chlorinity and salinity, the effect of dilution on the accidental spill of a soluble pollutant, and the concepts of steady state concentration and residence time.

Exercise 8, Temperature-Salinity-Density Relations and Density Stratification, investigates the relationships among temperature and density of fresh water and temperature, salinity, and density of seawater. Exercises include plotting the relationships between water density and depth at various seasons of the year in a fresh water lake, calculation of seawater salinity from analytical (weight) data, and a variety of exercises related to temperature-salinity-density diagrams.

Exercise 9, Coriolis Effect—Objects Moving on a Rotating Earth, helps students understand the Coriolis Effect. It includes work with examples in two dimensions (shooting a cannon at an enemy on a rotating turntable) and three dimensions (life on a rotating cylinder and life on the rotating earth).

Exercise 10, Atmospheric Circulation and Surface Ocean Currents, investigates the relationship between atmospheric circulation and ocean currents. Exercises consider the Hadley Cells of the atmosphere and their relationship to global climatic zonation, the Ekmann Spiral and its effects, the topography of the sea surface and its relationship to the speed and direction of surface currents, and the relationship between ocean circulation and biological productivity.

Exercise 11, Waves, provides an overview of the nature of ocean waves. The exercise covers orbital motion, depth of wave motion, fully developed sea, wave interference, wave velocity and group velocity, and the use of wave dispersion to determine how far the waves that break on the shore have traveled from the place they were generated.

Exercise 12, Wave Tank Experiments, provides the option of using an actual wave tank or using data provided to study the behavior of waves in a tank. The exercise emphasizes the determination of relationships among wavelength, wave period, and wave velocity, as well as the changes in wave motion that occur as deep-water waves move into shallow water and begin to "feel" bottom. For those with access to a wave tank, an exercise dealing with standing waves is included.

Exercise 13, Coastal Oceanography and Shoreline Erosion, presents some of the processes that affect the shores of oceans and large lakes. Students encounter longshore currents, longshore drift, and the effects of different settling rates for particles of different sizes. The effects of anthropogenic alteration of sediment sources and sediment transport, e.g., by damming of streams or building of structures along coasts are examined. Students also look at the different characteristics of emergent and submergent coastlines.

Exercise 14, The Carbon Cycle and the Greenhouse Effect, helps students develop a greater understanding of scientific issues related to anthropogenic carbon dioxide production. It provides an overview of the topic and especially the role of the oceans in regulating atmospheric carbon dioxide levels.

Exercise 15, Classification of Marine Organisms, introduces the student to the taxonomy of organisms. The exercise involves classifying 20 inanimate objects, and assigning organisms to the proper phylum based on their gross characteristics. Exercises on classification of organisms by habitat, mobility, and mode of nutrition are also included.

Exercise 16, Food Chains, Food Webs, and Biological Productivity, presents an overview of marine ecology and the potential of the world to feed growing human populations. Exercises focus on the conversion of solar energy to biomass, factors that limit biological production, iron as a limiting nutrient, the relation between nutrient concentration and oxygen concentration in the water column, the nitrogen cycle, trophic levels, food pyramids, food webs, and food chains.

Correspondence between Laboratory Exercises and Several Introductory Oceanography Textbooks

Exercise No.	Exercise Title	Thurman, *Introductory Oceanography* (7th Edition)	Thurman, *Essentials of Oceanography* (4th Edition)	Gross, *Oceanography* (6th Edition)	Duxbury and Duxbury, *An Introduction to the World's Oceans* (4th Edition)	Pinet, *Oceanography*	Garrison *Oceanography, an Introduction to Marine Science*
1.	The Earth and the Oceans—Basic Concepts	Chap. 1	Chap. 1	Chap. 2	Prologue	Chap. 1	Chap. 1
2.	Maps, Charts, and Marine Geography	Chap. 2	Chap. 2	Chap. 2	Chap. 1	Chaps. 1, 2	Chaps. 1, 2
3.	Bathymetry, Charts, and Navigation	Chap. 4	Chap. 2		Chap. 2	Chap. 2	Chap. 2
4.	Magnetics and Plate Motion Reconstructions	Chap. 3	Chap. 3	Chap. 3	Chap. 3	Chap. 3	Chap. 3
5.	Submarine Topography, Isostasy, Seismicity, and Plates	Chap. 4	Chap. 3	Chap. 3	Chap. 3	Chap. 3	Chap. 4
6.	Ocean Sediments	Chap. 5	Chap. 4	Chap. 11	Chap. 2	Chap. 4	Chap. 5
7.	Chemical Aspects of Seawater	Chap. 6	Chap. 5	Chap. 4	Chaps. 4, 5	Chap. 5	Chap. 6
8.	Temperature-Salinity-Density Relations and Density Stratification	Chap. 7	Chap. 6	Chap. 4	Chaps. 4, 5	Chap. 6	Chap. 7
9.	Coriolis Effect—Objects Moving on a Rotating Earth	Chaps. 7, 8	Chap. 6	Chap. 5	Chap. 7	Chap. 6	Chap. 8
10.	Atmospheric Circulation and Surface Ocean Currents	Chap. 8	Chap. 7	Chaps. 5, 7	Chaps. 6, 7	Chap. 6	Chap. 9
11.	Waves	Chap. 9	Chap. 8	Chap. 8	Chap. 8	Chap. 7	Chap. 10
12.	Wave Tank Experiments	Chap. 9	Chap. 8	Chap. 8	Chap. 8	Chap. 7	Chaps. 10, 11
13.	Coastal Oceanography and Shoreline Erosion	Chap. 11	Chap. 10	Chap. 10	Chaps. 10, 11	Chap. 9	Chap. 12
14.	The Carbon Cycle and the Greenhouse Effect	Chaps. 7, 8	Chaps. 6, 11	Chap. 5	Chap. 6	Chap. 14	Chap. 20
15.	Classification of Marine Organisms	Chap. 13	Chap. 12	Chaps. 13, 14, 15	Chap. 15, 17	Chap. 10	Chaps. 14–16
16.	Food Chains, Food Webs, and Biological Productivity	Chap. 14	Chap. 13	Chap. 12	Chap. 13	Chap. 11	Chap. 14

Contents

Contents

Name: ______________________ Date: ____________ Section: __________

Exercise 1
THE EARTH AND THE OCEANS—BASIC CONCEPTS

Purpose: You know many very fundamental things about the earth and the oceans: the size and shape of the earth, the fact that most of the earth is covered by oceans, the fact that the earth's interior is very different from its outer "skin" or crust. But it is likely that you have not thought very much about where these bits of common knowledge come from. In this exercise you will examine some of the evidence in support of our fundamental concepts.

Background required: No previous background is required for this exercise.

The metric system: In most of the world, physical measurements are made using the ***metric system***. The basic unit of distance is the ***meter***, abbreviated *m*. Some useful conversion factors are:

METRIC—ENGLISH CONVERSIONS
1 meter (m) = 39.37 inches
1 kilometer (km) = 1000 m (0.621 mile)
1 centimeter (cm) = .01 m (approx. 3/8 inch)
1 mile = 1.6 km
1 inch = 2.54 cm
1 gram (gm) = .035 ounce
1 kilogram (kg) = 1000 gm (2.2 pound)
1 ounce = 28.35 gm
1 pound = 453 gm (0.453 kg)

The basic unit of mass is the ***gram***. It is defined as the mass of a volume of water in a cube 1 cm on each side (strictly speaking, at approximately 4°C, but that detail need not concern us in this exercise. The change in the volume, with temperature, of a gram of water is relatively small).

The size and shape of the earth: We know from many lines of evidence that the earth is spheroidal (nearly spherical). Today it seems obvious, if only because we have all seen photographs of the earth taken from outer space. In earlier times the evidence was less blatant but no less clear-cut. It seems curious, therefore, that in much of the world for many years the earth was considered to be flat. This is perhaps a good example of the fact that what we see and how we interpret it is conditioned to a large extent by what we expect to see.

One bit of evidence for a spherical earth is the changing appearance of a ship as it sails toward the horizon (Figure 1.1). Rather than just getting smaller, as it would on a flat earth, we first lose sight of the lower parts of the ship. As it gets more distant, more and more of the ship slips below the horizon. The top of the mast is the last part to disappear from sight.

Another bit of evidence for a spherical earth comes from the differences in the elevations of stars (or the sun) measured at different locations on the earth at the same time (Figure 1.2). The ***North Star*** is positioned almost directly above the North Pole. The angle between the horizontal, an observer, and a star is called the ***elevation***, so we would say that at the North Pole, the North Star has an elevation of 90° (Figure 1.3). The further we go from the North Pole, however, the lower the elevation of the North Star, until at the equator it has an elevation of 0°. (In the Southern Hemisphere, we cannot see it at all.) The simplest interpretation of this set of observations is that the North Star is extremely far away from the earth and that the earth has a spherical shape.

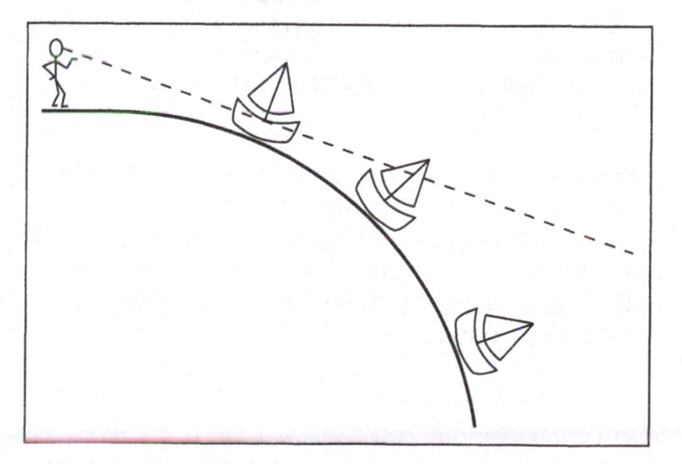

Figure 1.1. As a ship sails away, the part closest to the water disappears from view first. The tallest part of the ship is the last to vanish below the horizon.

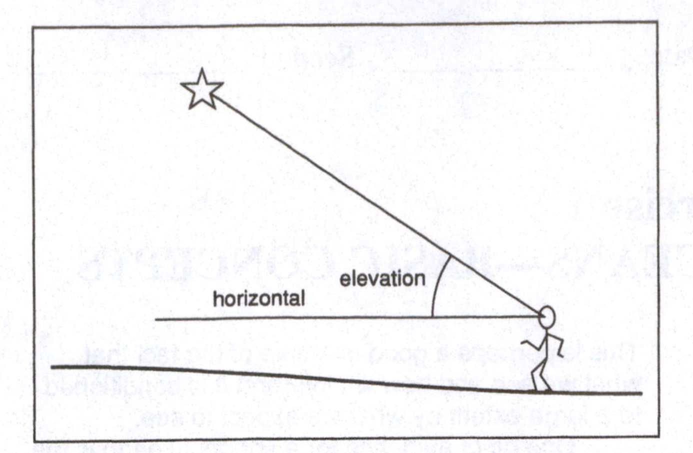

Figure 1.2. The elevation of a star is the angle that a line from the observer to the star makes with a horizontal line. The elevation of a star directly overhead is 90°. A star on the horizon has an elevation of 0°.

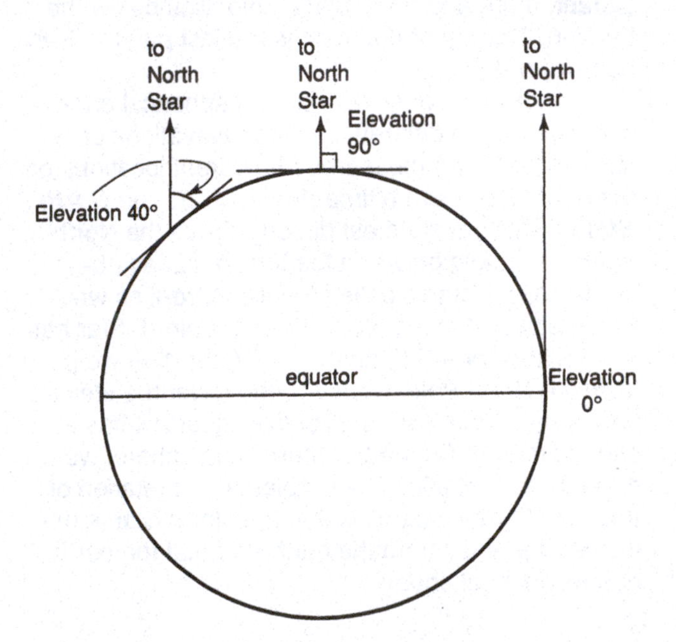

Figure 1.3. The North Star is almost directly above the North Pole. That means that an observer at the Pole measures its elevation to be 90°. An observer at the Equator measures its elevation to be 0°.

Problem 1: Explain why the conclusion drawn from the observations of the North Star that the earth is spheroidal requires that the North Star be extremely far away from the earth. How might the interpretation change if the North Star were just a few thousand miles from the earth?

Problem 2: Most of our discussion of the shape of the earth has considered only two choices, flat or spherical. But we can conjure up many different sorts of solid shapes in our minds: donuts, cylinders, tetrahedra, etc. (Figure 1.4). Imagine that the earth is the shape of a cube. Imagine, further, that you are standing at a place on a flat surface of that cube where the North Star is directly overhead. This point is labeled *starting point* on Figure 1.5. You follow the path shown on the diagram from the starting point to the *ending point* on the opposite side of the cube, making measurements of the elevation of the North Star as you go. Assuming the North Star to be very far away, describe how the measured el-

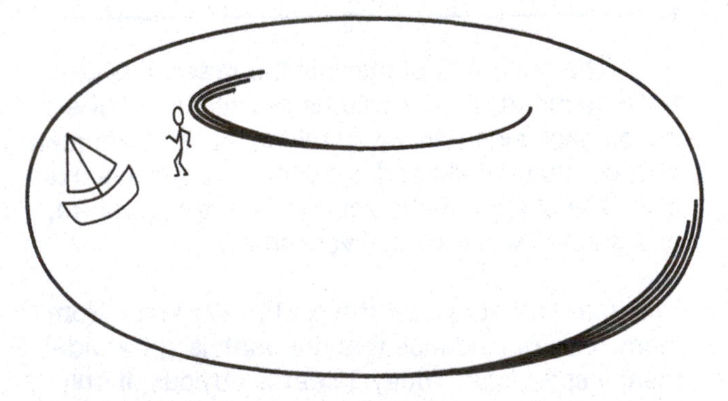

Figure 1.4. A sphere is not the only shape on which an observer would notice the lower parts of the ship disappearing first, as the ship sailed toward the horizon. A ship sailing in most directions away from an observer on a hypothetical donut-shaped planet would give the same appearance.

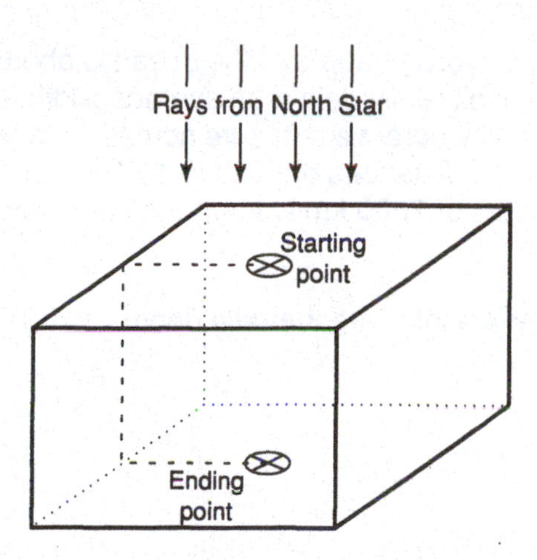

Figure 1.5. An observer on a hypothetical cubic earth travels along the path indicated by the thin dashed line.

evations of the North Star change as you take this journey.

By measuring the elevation of the North Star at two points along a north-south line on the surface of the earth, and measuring the distance between those two points, it is possible to calculate the circumference of the earth. Credit for the first measurement of this sort is often given to Eratosthenes (275–195 B.C.), a Greek who was the librarian of the famous library in Alexandria, Egypt. Eratosthenes noted that on one day each year the sun shone straight down a deep well in the city of Syene (now called Aswan). (It happened to be the Summer Solstice, but that is not essential to this measurement.) In other words, the elevation of the sun at noon in Aswan on that day was 90°. At *the same time* in Alexandria, which is approximately 800 km to the north, the elevation of the sun was 82.8° (Figure 1.6). (Noon is specified because at that time shadows are aligned in a north-south direction.)

For the reason shown in Figure 1.7, Angle ***a*** (the angle made by lines connecting the center of the earth with Aswan and Alexandria) is equal to Angle ***b***, the angle made by a vertical line and a line pointing to the sun at Alexandria.

Problem 3: Using Figures 1.6 through 1.8 as a guide, and remembering that there are 360° in a circle (and of course 90° in a *right angle*), answer the questions below.

How many degrees is angle ***a***, drawn from Alexandria to the center of the earth to Aswan? Explain your reasoning.

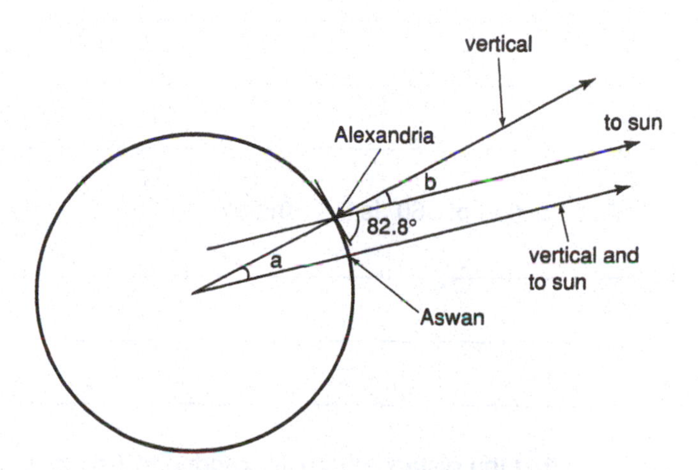

Figure 1.6. On the day Eratosthenes made his measurements, at noon, the sun was directly overhead at Aswan and was at an elevation of 82.8° at Alexandria. Because the sun is very far away from the earth, the lines drawn to the sun from both Alexandria and Aswan are parallel.

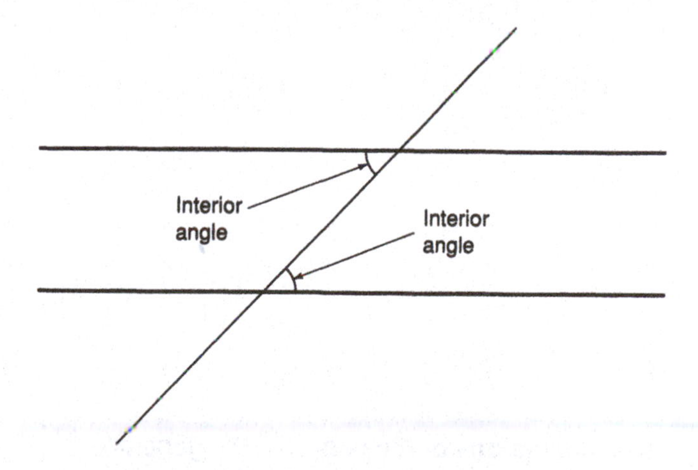

Figure 1.7. When two parallel straight lines are cut by a third straight line, the interior angles formed are equal (i.e., of the same size).

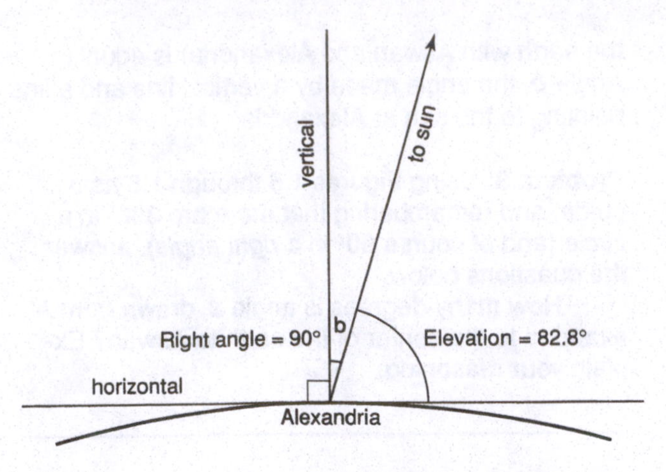

Figure 1.8. Close-up view of the relationship between the vertical and the direction to the sun at Alexandria.

____

____

____

What fraction of 360° is that angle?

____

____

____

Given that the distance from Alexandria to Aswan is 800 km, use the answer above to compute the circumference of the earth. Give your answer in km and in miles.

The dimensions of the oceans: We generally think of the oceans as being deep, but it is useful to have some feeling for the depth of the ocean relative to its horizontal dimensions.

The average depth of the ocean is about 3.8 km. We can't really define an average width, but depending on where we measure across oceans we typically get distances of 5000 to 10,000 km. Let's pick a width of 7600 km to keep the arithmetic simple.

Problem 4: Calculate the ratio depth:width.

Now imagine that we are trying to build a scale model of the ocean with the same depth:width ratio as the real ocean. Let's build our model in a pan that is 20 inches across. How deep should the water be in our model ocean? Show your calculations.

The mass and density of the earth: The mass of the earth can be estimated quite accurately from the way objects are attracted to it by the force of gravity. We will not go through the computations but will use a figure for the mass of the earth of 5.98×10^{27} gm. The radius of the earth is 6370 km (a figure that comes from more accurate measurements than those of Eratosthenes).

Note that the figure given for the earth's mass, 5.98×10^{27}, is expressed in scientific notation. *Scientific notation* is a convenient way to express very large and very small numbers. In scientific notation, a number is expressed as a small number (usually between 1 and 10) that is multiplied by 10 raised to the appropriate power. Remember that

$10^{-5} = .00001$
$10^{-4} = .0001$
$10^{-3} = .001$
$10^{-2} = .01$
$10^{-1} = .1$
$10^{0} = 1$
$10^{1} = 10$
$10^{2} = 100$
$10^{3} = 1,000$
$10^{4} = 10,000$
etc.

As an example, 433,000 can be conveniently written as:

$4.33 \times 100,000$

or

4.33×10^{5}

and 0.0000472 can be written as

4.72×0.00001

or

4.72×10^{-5}

Problem 5: What is the radius of the earth in cm? (Use scientific notation.)

What is the volume[1] of the earth in cubic centimeters (cm^3)? Continue to use the value of 6370 km for the radius of the earth, but remember to convert that value to cm in order to calculate the volume in cm^3. Hint: 1 km = 10^5 cm. Show your calculations.

Remembering that ***density*** is defined as mass per unit volume (or the mass of an object divided by its volume), calculate the average density of the earth.

The average density of the rocks at the surface of the earth is about 2.7 gm/cm^3. So if your answer to the last part of Problem 5 was correct, you realize that the average density of the earth is considerably greater than the average density of the rocks at the earth's surface. This of course tells us that the density of the material in the interior of the earth is considerably greater than the density of the material at the surface.

We are unable to make direct observations or take samples of the deep interior of the earth, so our information about what is down there must come from indirect evidence. Among the lines of evidence that geophysicists use to draw conclusions about the nature of the earth's interior is the density of the earth (and the way in which density varies with depth below the surface). In other words, calculations such as the one you just made in Problem 5 tell us very important things about the part of the earth we cannot examine directly.

---

[1] The *volume of a sphere* can be calculated using the formula

$$\text{Volume} = 4/3\ \pi r^3$$

where π is the ratio of the circumference to the diameter of a sphere and has the value 3.1416. r is the radius of the sphere.

Name: ______________________ Date: ____________ Section: ____________

Exercise 2
MAPS, CHARTS, AND MARINE GEOGRAPHY

Purpose: Scientists often speak of the ***world ocean***, stressing the fact that all of the oceans on the globe form a single interconnected body of water. Events and processes in the North Pacific may affect the conditions or the organisms in the South Atlantic or the Mediterranean. In order to fully appreciate the global scale of oceanographic processes it is necessary to be familiar with basic concepts of geography and the construction of maps and charts.

In this exercise you will review the concepts of latitude and longitude, and will acquire an appreciation of the problems of representing the spherical earth on a two-dimensional drawing (i.e., a piece of paper). You will also review your knowledge of the locations of important regions of the oceans and coastal areas.

Equipment needed: In addition to this laboratory manual, you will need to have access to a globe that shows lines of longitude and latitude.

Latitude and Longitude: We are all familiar with the use of coordinates to describe locations on a flat surface. The use of x- and y-coordinates on a piece of graph paper is certainly familiar to you, and you probably can plot a point like *x=5, y=7* on a graph with no difficulty. This system of describing locations on a flat surface is often called the Cartesian system, after the 18th century French mathematician, René Descartes. Many cities use this approach in the naming of streets. Visitors to Manhattan, for example, would have little problem finding their way from the corner of 3rd Avenue and 14th Street to the corner of 7th Avenue and 33rd Street. It is similarly straightforward to express any location on the surface of a sphere in terms of two coordinates. Locations on the surface of the earth are commonly expressed in terms of ***longitude*** and ***latitude***.

The latitude of any point on the surface of the earth is the angle between a line drawn from that point to the center of the earth and a plane through the equator (Figure 2.1). The latitude of the equator is therefore 0°, and the latitude of either pole is 90°. To avoid ambiguity, the latitude of any point between the equator and the North Pole is specified as **N (North)** and the latitude of any point between the equator and the South Pole is specified as **S (South)**.

The line of intersection between a sphere and a plane is always a circle. The circle formed by the intersection of a sphere and a plane that does not pass through its center is called a ***small circle***. Lines of equal latitude are small circles parallel to one another and to the equator (Figure 2.2).

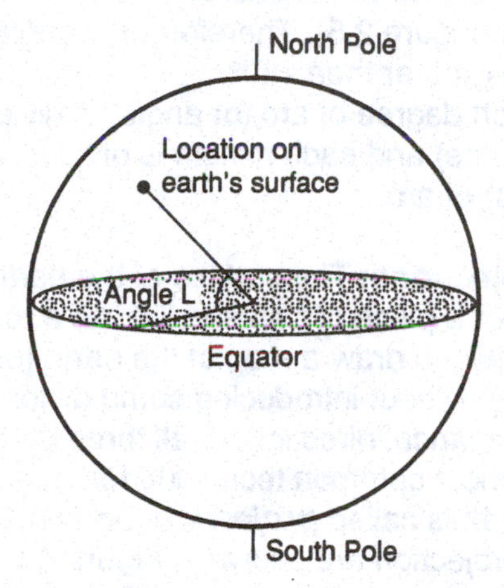

Figure 2.1. The latitude of any location on the surface of the earth is the angle (Angle L) between a line connecting that point to the center of the earth and a plane through the equator.

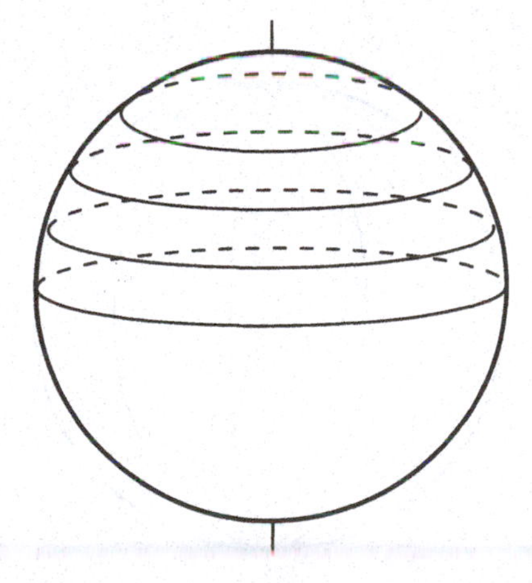

Figure 2.2. Lines of equal latitude are small circles drawn parallel to the equator. Dashed lines are on the side of the earth away from the observer.

Circles created by the intersection of the surface of a sphere with a plane that passes through the center of the sphere are called ***great circles***. Great circles that pass through the north and south poles are called ***meridians*** or ***lines of longitude*** (Figure 2.3).

The rotation of the earth about its axis makes the equator an obvious choice as a reference position relative to which latitudes (i.e., positions north or south) can be measured. There is no comparable obvious choice for a meridian relative to which position in the east-west direction can be referred. By convention in most of the world, longitudes are measured relative to the ***Prime Meridian*** in Greenwich, England. That is, the longitude of this meridian is designated as 0° (Figure 2.4).

The longitude of every other point on the earth is the angle between the Prime Meridian and the meridian on which that point lies. There are 360° in a circle, but by convention, longitudes are measured as angles to the east or west of the Prime Meridian (Figure 2.5). Therefore, no longitude has an angle greater than 180°.

Each degree of arc (or angle) is divided into 60′ (minutes) and each minute is divided into 60″ (seconds) of arc.

Map projections: The surface of the earth is curved but a piece of paper is flat. As a result, it is not possible to draw a map of the earth (or a part of the earth) without introducing some distortion of shape, distance, direction, or all three of these.

The most common technique for creating a map of the earth is called ***projection***. Some ways of creating a projection are shown in Figure 2.6 Imagine that the earth is transparent and that there is a bright light at its center. The light rays shine through to a screen, projecting onto the screen the outlines of continents, ocean, rivers, and any other features

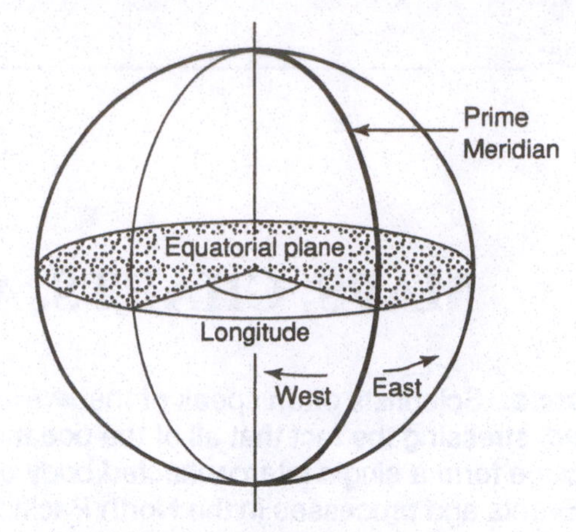

Figure 2.4. The longitude of a point is the angle between the meridian on which the point lies and the Prime Meridian.

of interest. The screen itself may be any shape that can be constructed from a flat piece of paper. Commonly used projections are flat, cylindrical, and conic. If the screen is not flat, it can be unfolded or unrolled to form a flat image, or map.

There are a number of ways to draw a map, and the kind and amount of distortion produced by each is different. Distortion is normally least in regions where the screen is tangent to the globe and is greatest where the screen is furthest from the globe. In a cylindrical projection, the regions furthest from the equator are enormously distorted. The ***Mercator projection*** (Figure 2.7) is a modification of a simple cylindrical projection. Distortion on this map is significant at high latitudes, but it is less than in the simple cylindrical projection. At any point on the Mercator projection the scale in the east-west direction is the same as the scale in the north-south direction. In addition, because longitude and latitude lines are everywhere parallel to one another, the

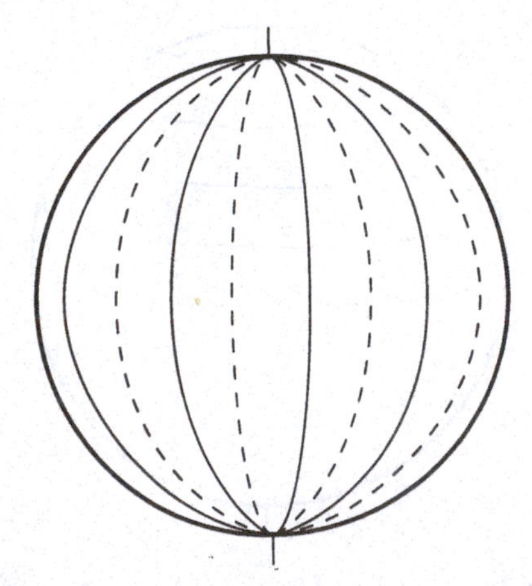

Figure 2.3. Meridians, or lines of longitude, are great circles that pass through the North and South Poles. Dashed lines are on the side of the earth away from the observer.

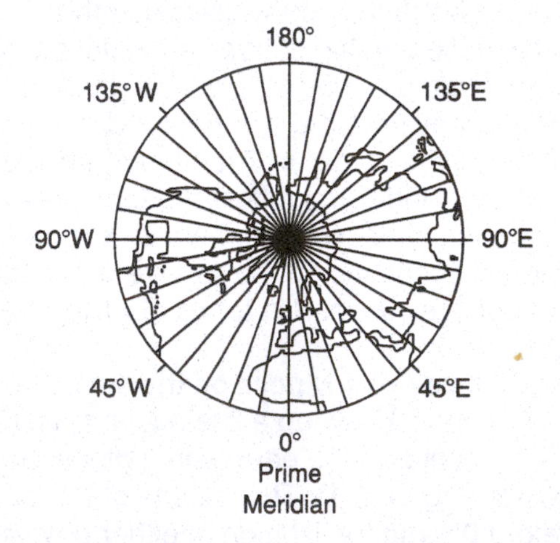

Figure 2.5. The world as viewed looking downward from a point in space above the North Pole. Longitudes are measured as degrees east or west of the Prime Meridian.

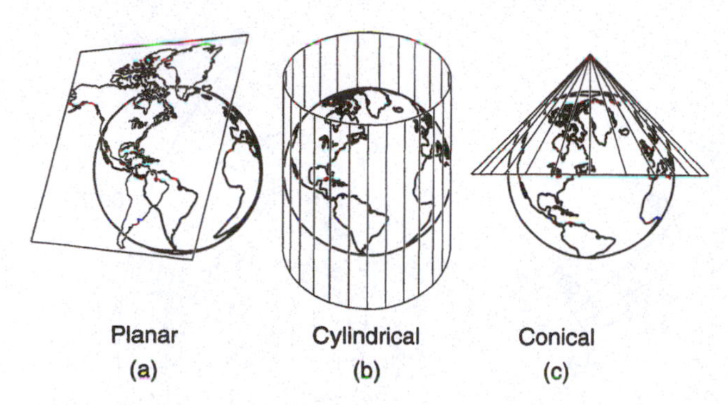

Figure 2.6. Generation of planar, cylindrical and conic map projections. (Adapted from U.S. Navy Hydrographic Office Pub. No. 9, figs. 304, 314, and 317a, 1966.)

course of a ship steering a constant compass direction plots as a straight line, called a ***rhumbline***, on a Mercator projection. The shortest distance between two points on the surface of the earth is a great circle. Except when the two points lie on the equator or along the same meridian, the great circle course connecting them will appear on the Mercator projection as a curved line.

A ***conic projection*** is shown in Figure 2.8. In non-equatorial regions the conic projection is more accurate than a Mercator projection. A variant is the ***polyconic projection***, which consists of a series of conic projections pieced together along small circles (Figure 2.9). Because the screen in the polyconic projection is tangent to the surface of the earth along more than one small circle, it depicts the earth's surface with less distortion than a simple conic projection or a Mercator projection.

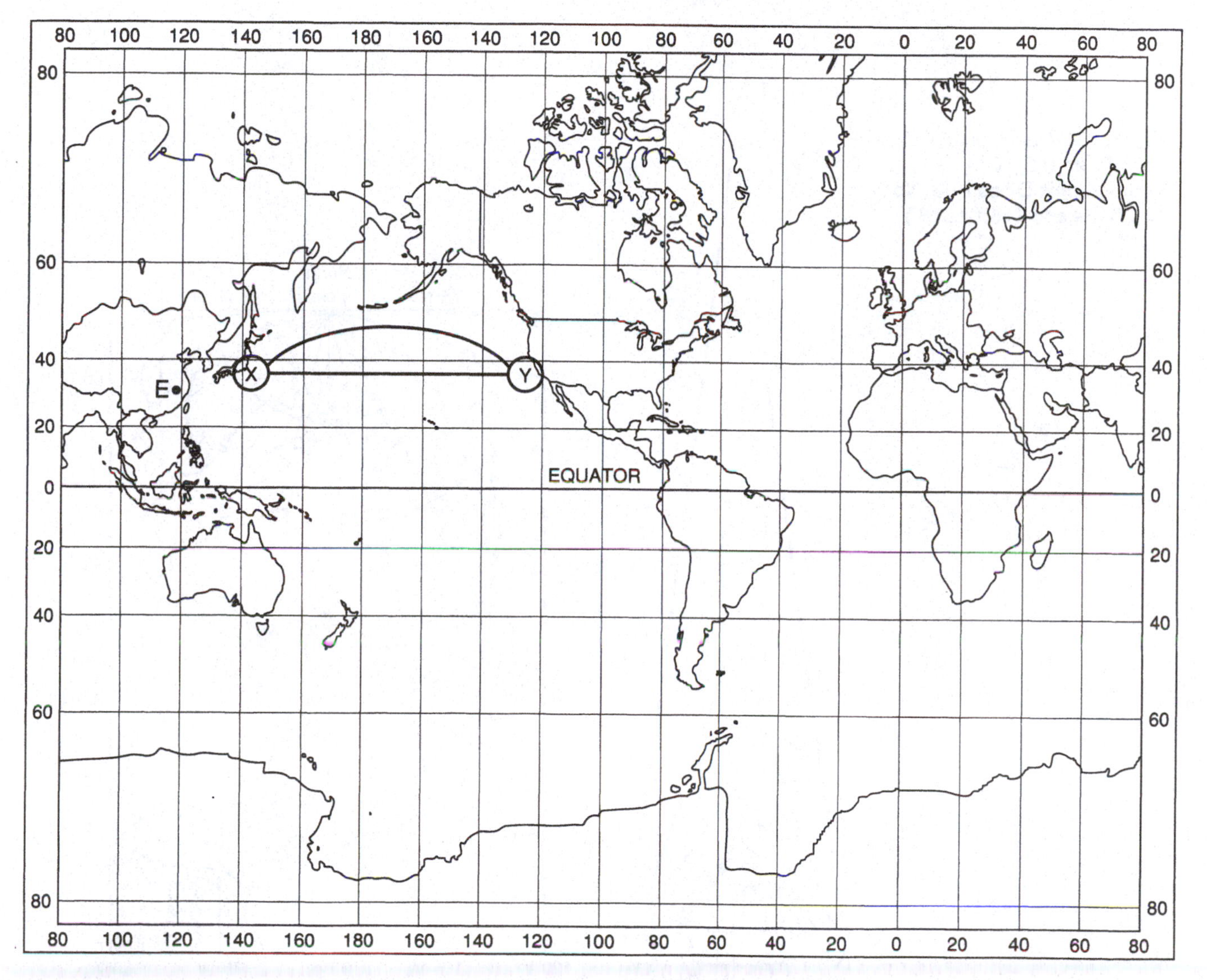

Figure 2.7. Mercator projection of the world. The straight line connecting points **X** and **Y** is called a rhumbline. It represents the course of a ship steering a constant compass direction from **X** to **Y**. The shortest distance between **X** and **Y**, the great circle route, is the curved line connecting the points. (Adapted from U.S. Navy Hydrographic Office Pub. No. 9, fig. 305, 1966.)

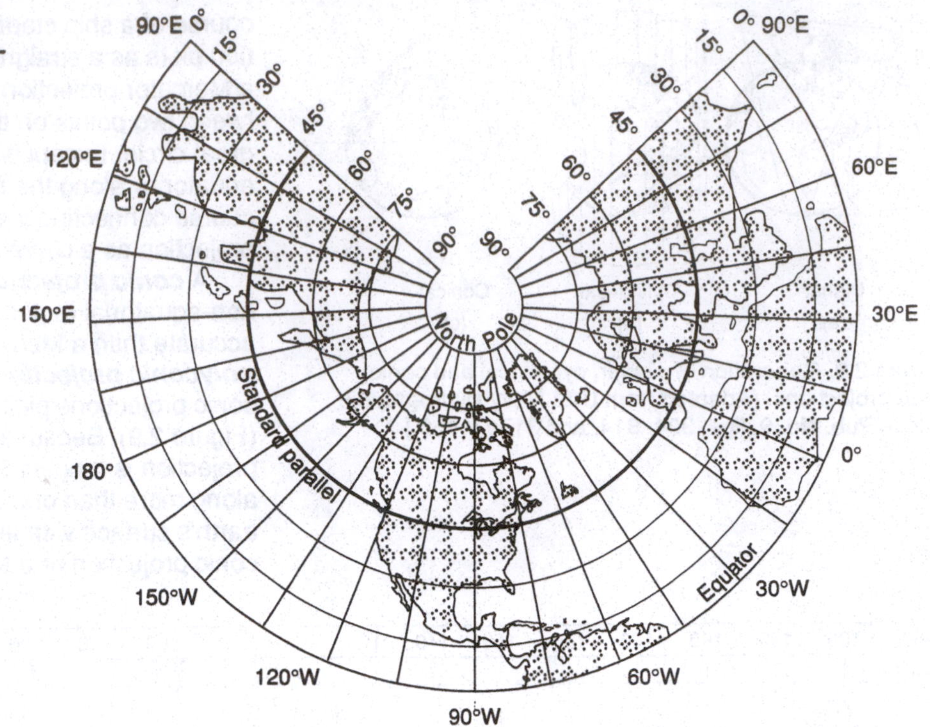

Figure 2.8. A conic projection. (Adapted from U.S. Navy Hydrographic Office Pub. No. 9, Fig. 313b, 1996.)

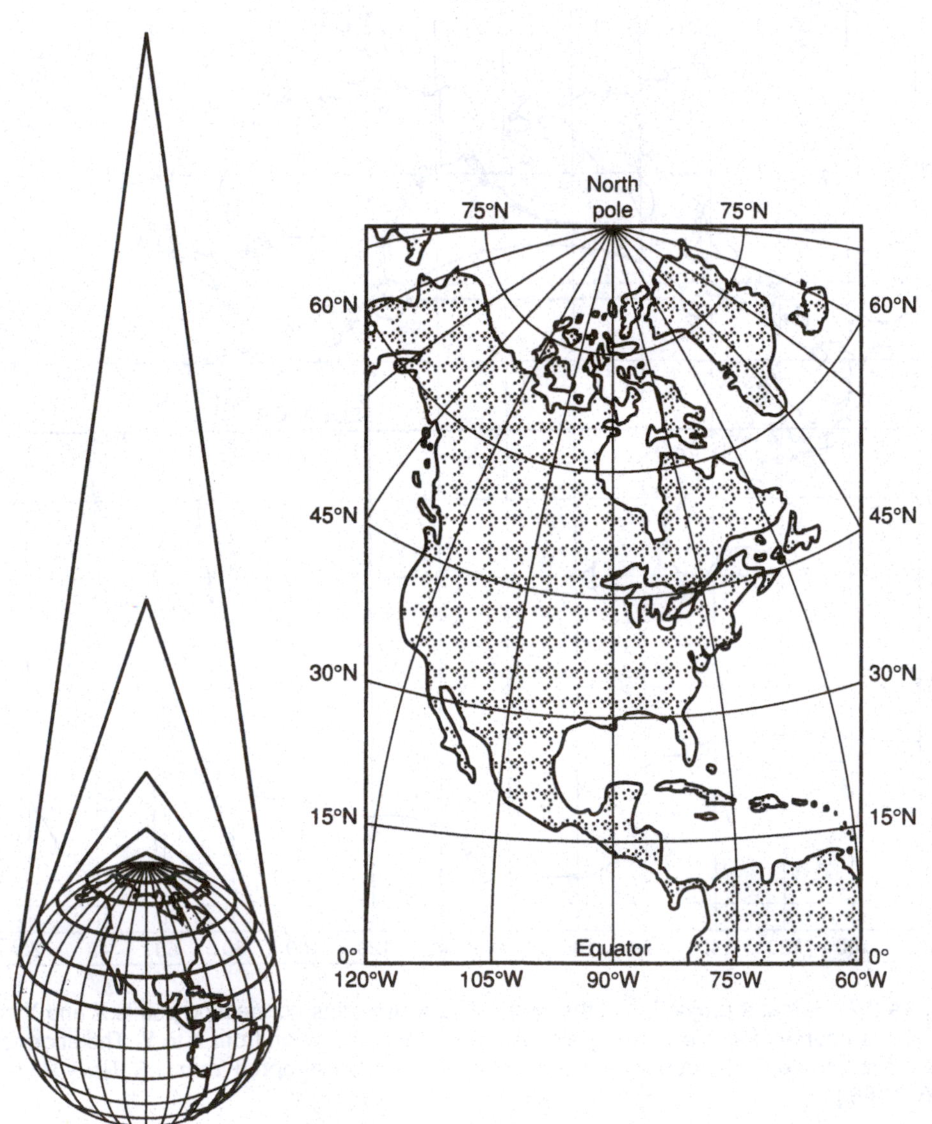

Figure 2.9. A polyconic projection. (From U.S. Navy Hydrographic Office Pub. No. 9, figs. 315a and 315b, 1966.)

Problem 1: On the Mercator projection in Figure 2.7, plot the points with the following latitudes and longitudes:

Point	Latitude	Longitude
A	45°N	45°W
B	15°S	105°E
C	0°N	0°W
D	60°S	75°W

Problem 2: The letter **E** is plotted on the Mercator projection within China. Calculate the longitude and latitude of the point exactly on the opposite side of the earth (i.e., antipodal) to point **E**. It is the point from which you could dig straight down to reach China. Mark that point on the map with the letter **F**.

Problem 2 work space:

Another very useful projection is shown in Figure 2.10. In this projection, the distortion of both shapes and areas is minimized, at the expense of splitting the world in several places. We will be using projections that split the continents and leave the oceans largely uncut. But the splits could equally well have been placed to cut the oceans and leave the continents whole.

Problem 3: Consider the map in Figure 2.10. For each 10° band of latitude estimate the percentage of the surface that is water and the percentage that is land. (Remember that for each 10° band the two percentages must add up to 100.) Enter your results in Table 2.1.

Problem 4: Using your answers in Table 2.1 as a guide, describe in a few sentences the difference between the distribution of land and water in the Northern Hemisphere and in the Southern Hemisphere.

The Northern Hemisphere has a higher percentage of land than the Southern Hemisphere, which has more ocean than land

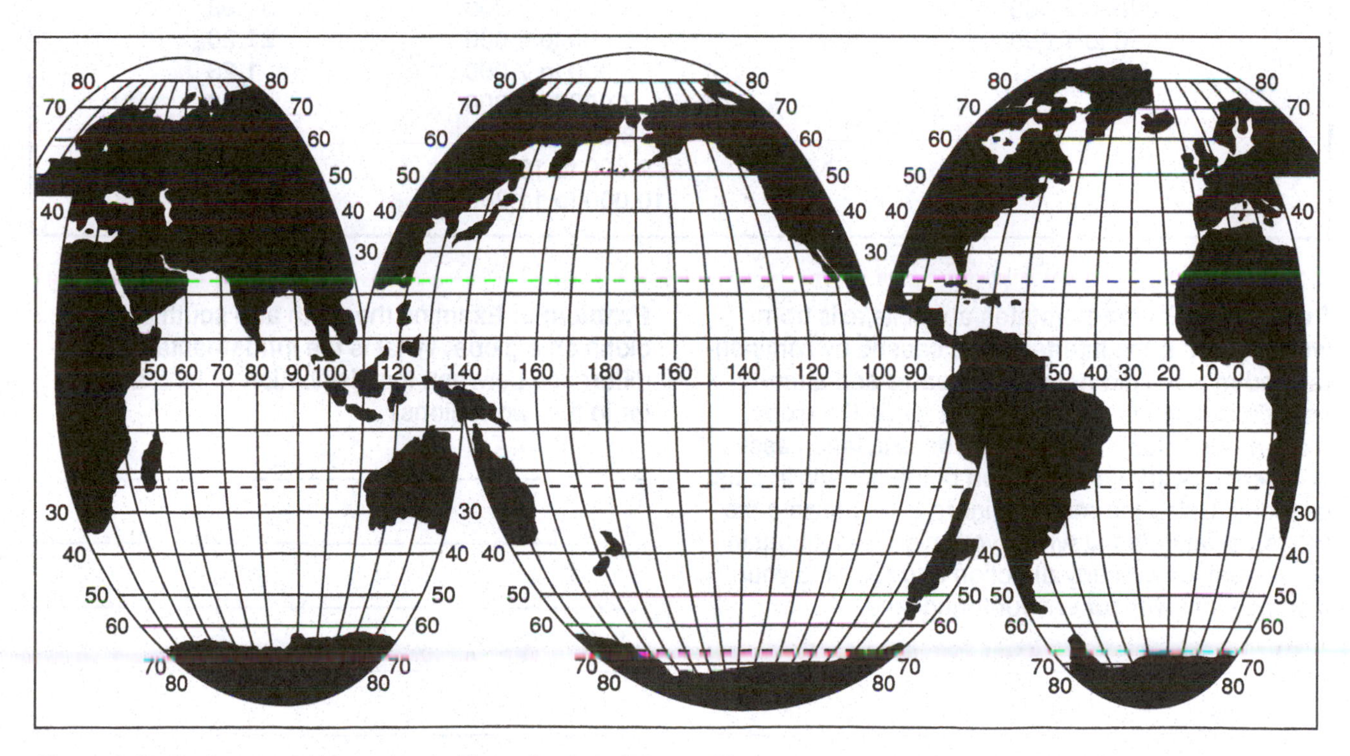

Figure 2.10. An interrupted homolographic projection of the world.

Table 2.1. Global Distribution of Land and Water. Write answers to Problem 3 here.

Range of Latitudes	Percent Water	Percent Land	Total	Range of Latitudes	Percent Water	Percent Land	Total
70° to 80°N	65	35	100	0° to 10°S	75	25	100
60° to 70°N	10	90	100	10° to 20°S	67	33	100
50° to 60°N	45	55	100	20° to 30°S	70	30	100
40° to 50°N	40	60	100	30° to 40°S	70	30	100
30° to 40°N	51	49	100	40° to 50°S	95	5	100
20° to 30°N	57	43	100	50° to 60°S	97	3	100
10° to 20°N	67	33	100	60° to 70°S	89	11	100
0° to 10°N	75	25	100	70° to 80°S	1	99	100

Table 2.2. Distribution of the earth's surface as a function of elevation above or below sea level.

Elevation above sea level (meters)	Percent of earth's surface	Elevation below sea level (meters)	Percent of earth's surface
Above 5,000	0.1	0 to 1,000	11.91
4,000 to 5,000	0.4	1,000 to 2,000	4.38
3,000 to 4,000	1.1	2,000 to 3,000	8.50
2,000 to 3,000	2.2	3,000 to 4,000	20.94
1,000 to 2,000	4.5	4,000 to 5,000	31.69
0 to 1,000	20.8	5,000 to 6,000	21.20
		6,000 to 7,000	1.23
		7,000 to 8,000	0.10
		8,000 to 9,000	0.03
		9,000 to 10,000	0.01
		10,000 to 11,000	0.00

Problem 5: The earth rotates about its axis from west to east. If the interferences caused by continental land masses did not exist, oceanic and atmospheric circulation patterns might girdle the globe, more or less along lines of latitude. But landmasses do interfere with the circulation of the oceans at most latitudes. Using the map in Figure 2.10, indicate the ranges of latitudes at which you might sail a boat in an easterly or westerly direction and return to your starting point without encountering land.

Traveling along the 60°S latitude will allow you to travel around the globe w/o hitting land

Problem 6: Examine the north and south polar regions on a globe. What is the fundamental difference between the distribution of land and water in the two regions?

The Northern Polar region is mostly water. The Southern Polar region is mostly land.

Figure 2.11. Plot answer to Problem 7 here.

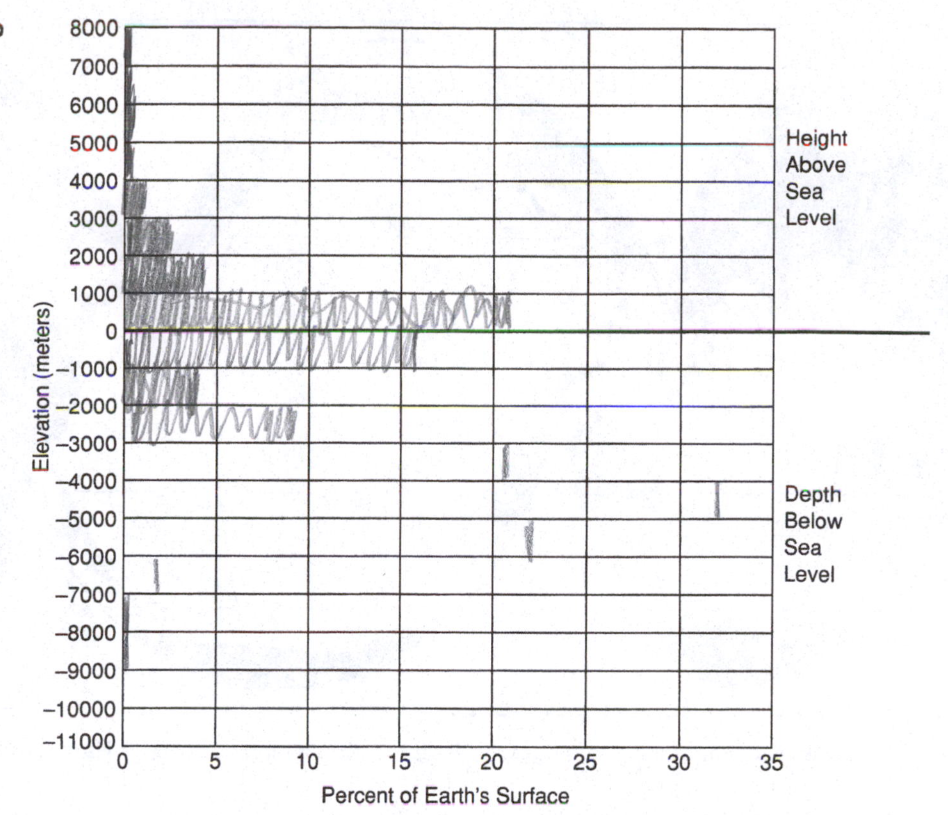

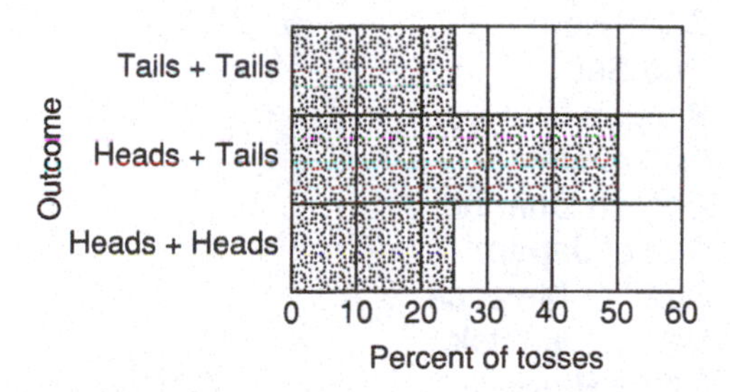

Figure 2.12. Histogram showing the expected results of a large number of coin tosses.

The hypsographic curve: An important aspect of global geography is the distribution of elevations, or ***hypsography***. Table 2.2 is a compilation of the percentages of the earth's surface in each thousand meter (above or below sea level) band of elevation.

Problem 7: Using the data in Table 2.2 and the axes of Figure 2.11, plot a ***histogram*** of the distribution of elevations of the earth's surface. A histogram is a type of bar graph that shows the percentages of some measurement that are within a series of specified ranges. For example, imagine that you have several hundred quarters, and you flip each of them twice. You will find that in about 25 percent of the cases you will get heads twice. In about another 25 percent of the cases you will get tails twice. And approximately 50 percent of the time you will get one head and one tail. This is illustrated in the histogram in Figure 2.12.

A histogram with only a single peak or maximum, such as the histogram of coin tosses in Figure 2.12, is said to be ***unimodal***. A unimodal distribution may often describe a single population or be the result of a single process. A histogram with two peaks is said to be ***bimodal***, and may often describe the combination of two different populations or be the result of two different processes. A histogram with more than two peaks is ***polymodal***. Describe the distribution of elevations in the histogram you drew. What can you conclude about the number of different kinds of terrains represented or the number of processes responsible for the global distribution of elevations?

Most of the Earths surface that is above the ocean is between 0-1,000 meters above sea level. Most of the Earths surface beneath the ocean is between 3,000-6,000 meters below sea-level.

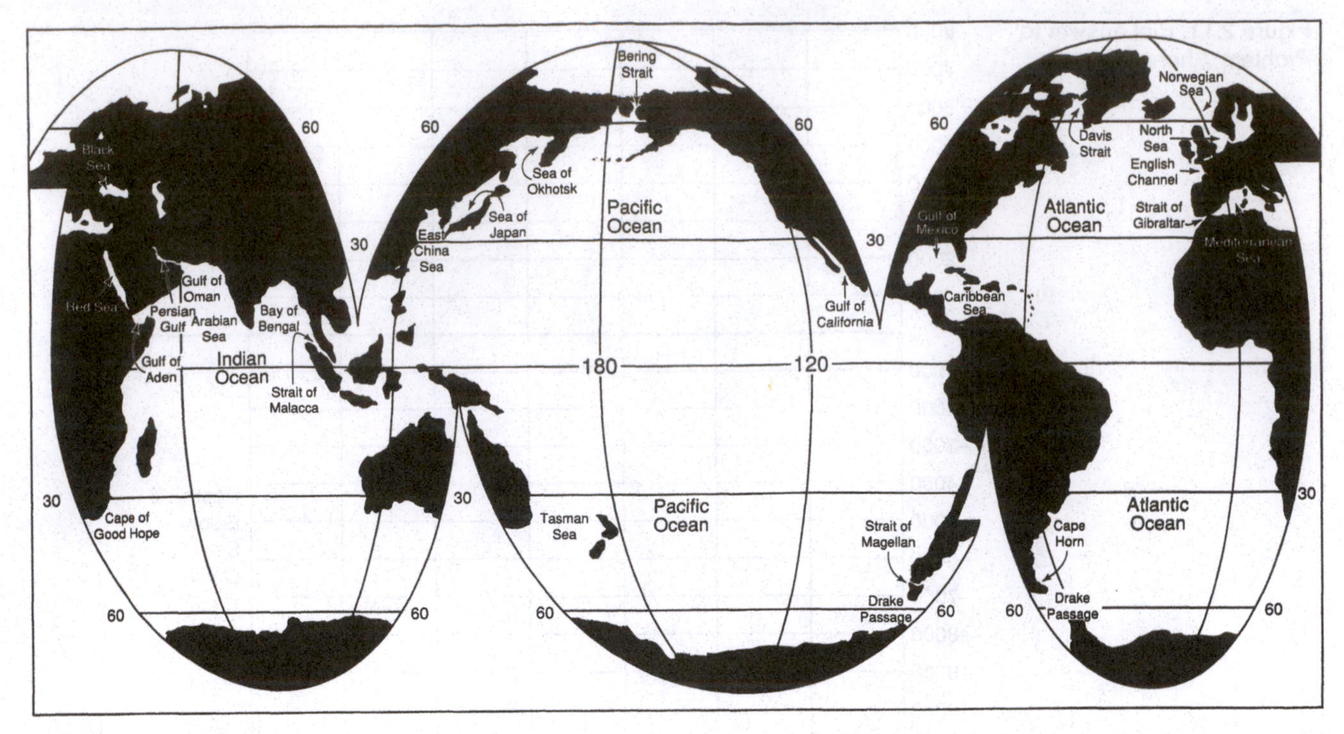

Figure 2.13. Locations of important oceanographic features. (Base map courtesy of National Ocean Service.)

Important locations in and around the oceans. Figure 2.13 is a map of the world on which a number of important geographic features are indicated. Note that the term ***ocean*** is used to refer to large sections of the global ocean. The term ***sea*** is used to refer to a smaller body of water often partially isolated from the open ocean by land. With the exceptions of the Black Sea and the Mediterranean Sea and its subdivisions, seas are regions within oceans. The Arctic Ocean is in some respects better thought of as a sea. Although it is large, its connections with the global ocean are somewhat restricted.

Problem 8: Indicate on the map in Figure 2.14 the locations of the following places. Use the letters preceding the place name where there is not enough space to write the entire name.

a. Atlantic Ocean
b. Pacific Ocean
c. Indian Ocean
d. Caribbean Sea
e. Mediterranean Sea
f. Gulf of California (Sea of Cortez)
g. Gulf of Mexico
h. Black Sea
i. Bering Strait
j. Strait of Magellan
k. Strait of Gibraltar
l. Cape of Good Hope
m. Cape Horn
n. Red Sea
o. Persian Gulf
p. Gulf of Oman
q. English Channel
r. Sea of Japan
s. Strait of Malacca
t. Sea of Okhotsk
u. Drake Passage
v. Norwegian Sea
w. Gulf of Aden
x. Bay of Bengal
z. Arabian Sea
aa. North Sea
bb. Davis Strait
cc. Tasman Sea
dd. East China Sea

Problem 9: Look at the map in Figure 2.13. Of the three major oceans of the world (Atlantic, Pacific, and Indian), two extend from high northern latitudes to high southern latitudes, and one does not. Which of the three oceans has only one polar extreme?

____

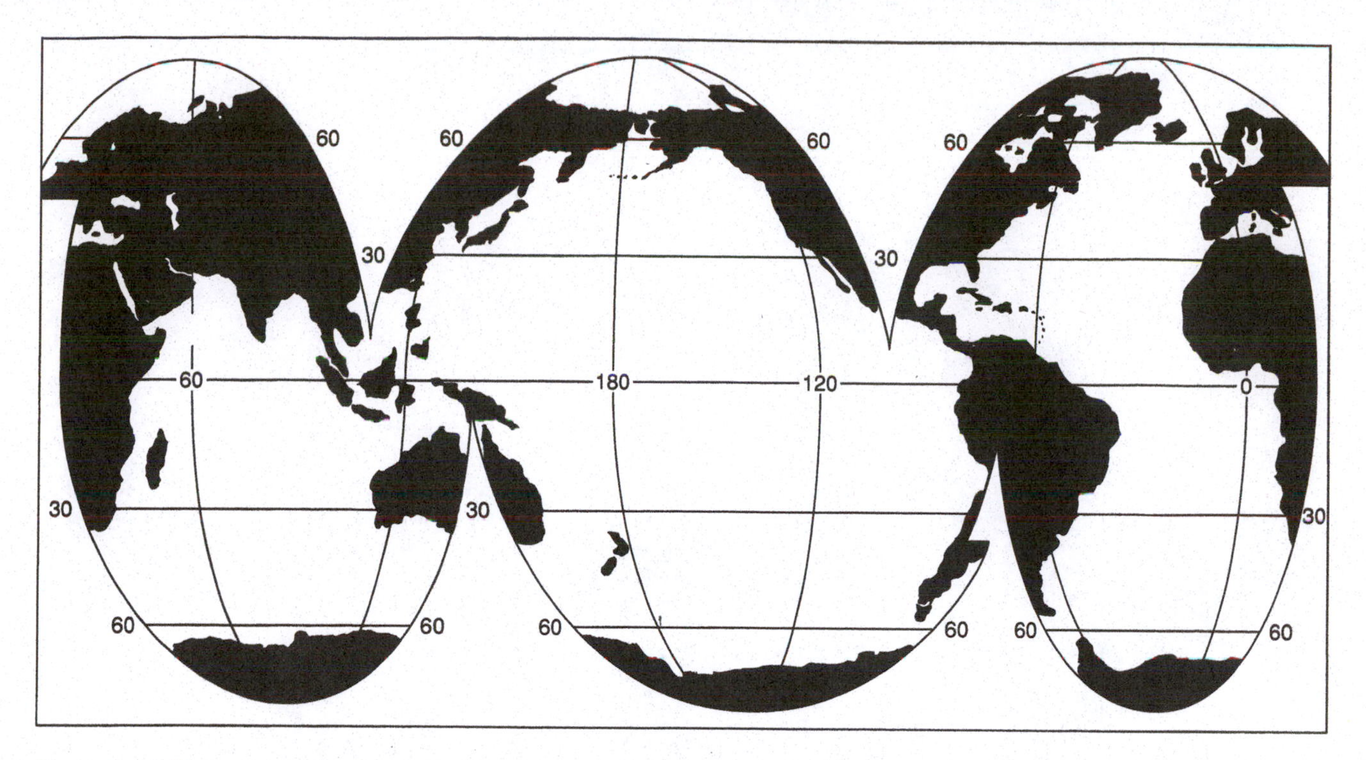

Figure 2.14. Plot answers for Problem 8 here. (Base map courtesy of National Ocean Service.)

Name: ________________________ Date: ____________ Section: ____________

Exercise 3
BATHYMETRY, CHARTS, AND NAVIGATION

Purpose: Charts are among the basic tools of navigators as well as oceanographers. They may contain a wealth of information about water depths, navigational aids, and the location of geographic features. In this exercise you will gain experience in reading and using the information on marine charts. You will develop some facility in drawing and interpreting depth contours and in relating map views of underwater land forms to cross-sectional views.

Equipment needed: You will need a supply of sharp pencils, a good eraser, a straightedge, and a set of parallel rulers (or a square of cardboard and a scissors).

Types of charts: Maps of the oceans (and inland waterways) used by oceanographers, limnologists, and mariners typically show a great deal of information in addition to the location of shorelines and geographic features. Such maps are called ***charts***. ***Hydrography*** is the charting of oceans, lakes, and rivers, especially for navigational purposes. Hydrographic charts are the most common and readily accessible way of presenting and obtaining hydrographic information.[1] An example which we will use in this exercise, a hydrographic chart of the West End of Lake Erie, is shown in Plate I.

The measurement of water depths is called ***bathymetry***, and a chart that displays bathymetric data is a ***bathymetric chart***.[2]

Measuring distances and speeds: In most of the world distances are measured in the metric system. Ship velocities are usually given in kilometers per hour (km/hr) while velocities of oceanic currents might be quoted in centimeters per second (cm/sec). Some useful conversion factors are:

1 km	=	1000 m
1 m	=	100 cm
1 inch	=	2.54 cm
1 m	=	39.37 inches
1 km	=	.6214 mile
1 mile	=	1.609 km

In the United States, nautical distances are frequently measured in ***nautical miles*** (6076 feet) rather than ***statute miles*** (5280 feet), although conversion to the metric system is occurring slowly. A nautical mile is almost, but not quite, equal to 1/60 of a degree (0.0167°) or one minute (1′) of arc at the earth's equator. (In the Great Lakes and other inland waterways, however, it has long been customary to measure distances in statute miles.) When distances are measured in nautical miles, velocities are given in nautical miles per hour, or ***knots***. It is incorrect to speak of velocities as "knots per hour."

Problem 1: As you do this problem don't just grind out the answers, but look at them and try to develop some feeling for the relationships among values given in different units.

1a.) The Gulf Stream flows at an average velocity of 100 centimeters per second. Before doing any computations, do you think that this is faster or slower than you walk? Check one.

Faster ✓ Slower ______ Same ______

1b.) Calculate the average velocity of the Gulf Stream in units of:

knots 1.94

100cm | 1in / 2.54cm | 1ft / 12in | 3.28 ft/cm / 6.076

= 0.00054 × 60 × 60

---

[1] Hydrographic charts are published by many government agencies, worldwide. In the United States they are available directly from

Distribution Branch (N/CG33)
National Ocean Service
Riverdale, Maryland 20737-1199
Phone (301) 436-6990

and are also sold in many yachting supply stores.

[2] While a hydrographic chart often contains bathymetric data, a bathymetric chart need not show navigational information.

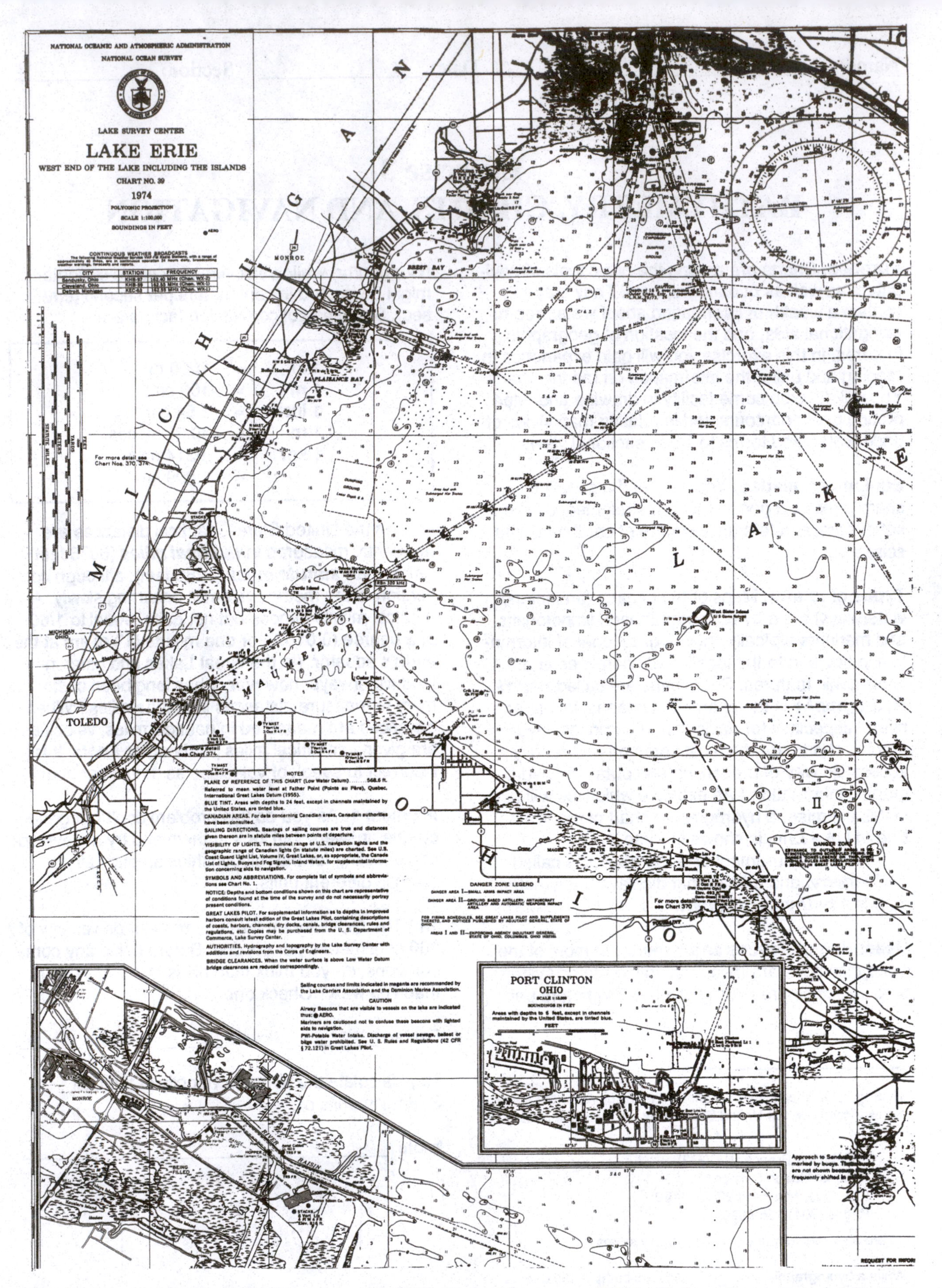

Plate I: West end of Lake Erie. (Reproduced from U.S. Dept. of Commerce, NOAA—National Ocean Survey, Lake Survey Center Chart L.S. 39.)

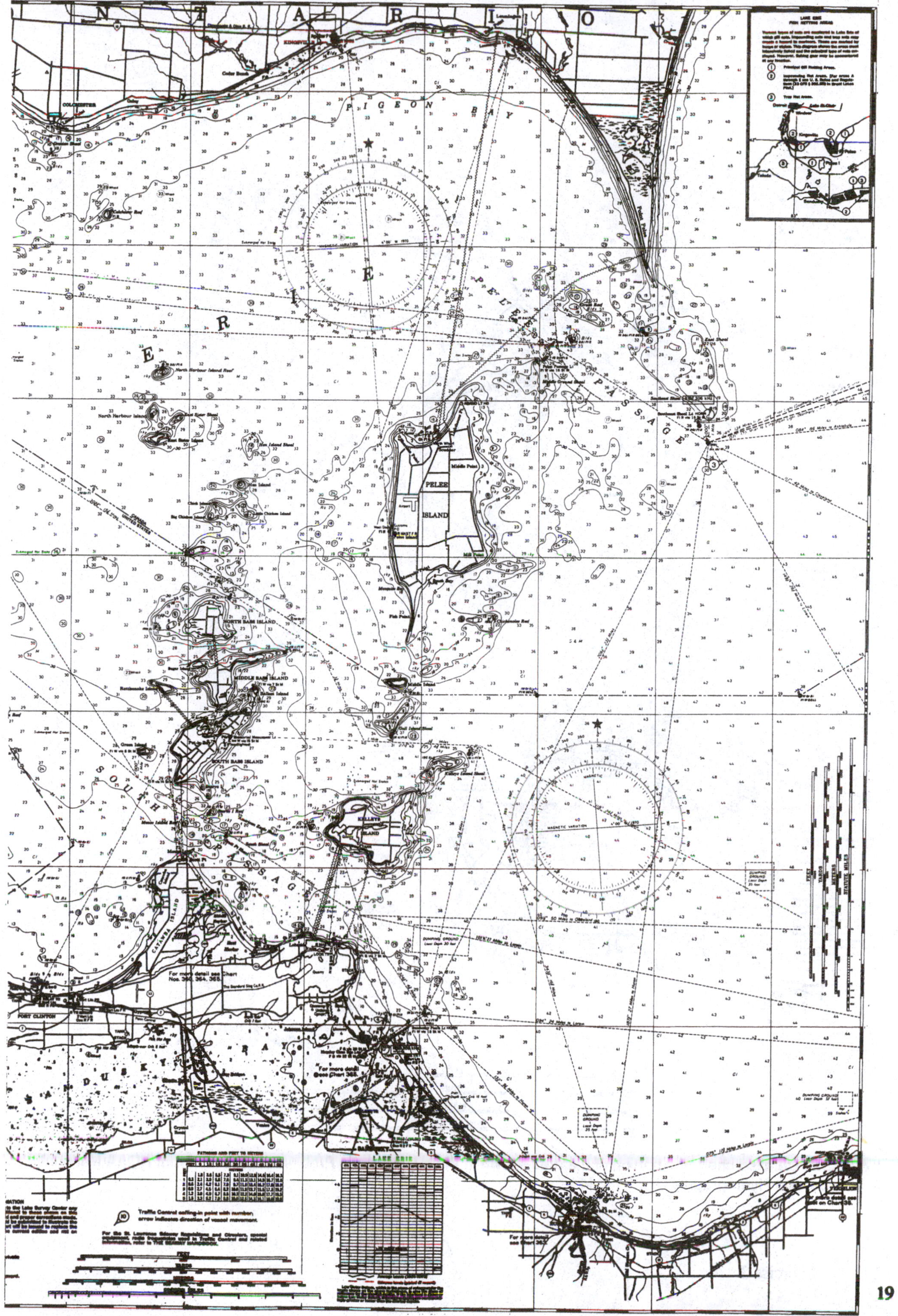

PIGEON BAY
PELEE PASSAGE
PELEE ISLAND
NORTH BASS ISLAND
MIDDLE BASS ISLAND
SOUTH BASS ISLAND
KELLEYS ISLAND
SOUTH PASSAGE
SANDUSKY BAY
PORT CLINTON
North Harbour Island
Middle Island
Colchester
Kingsville
MAGNETIC VARIATION
For more detail see Chart Nos. 360, 364, 365.
DUMPING GROUND
Traffic Control calling-in point with number, arrow indicates direction of vessel movement.

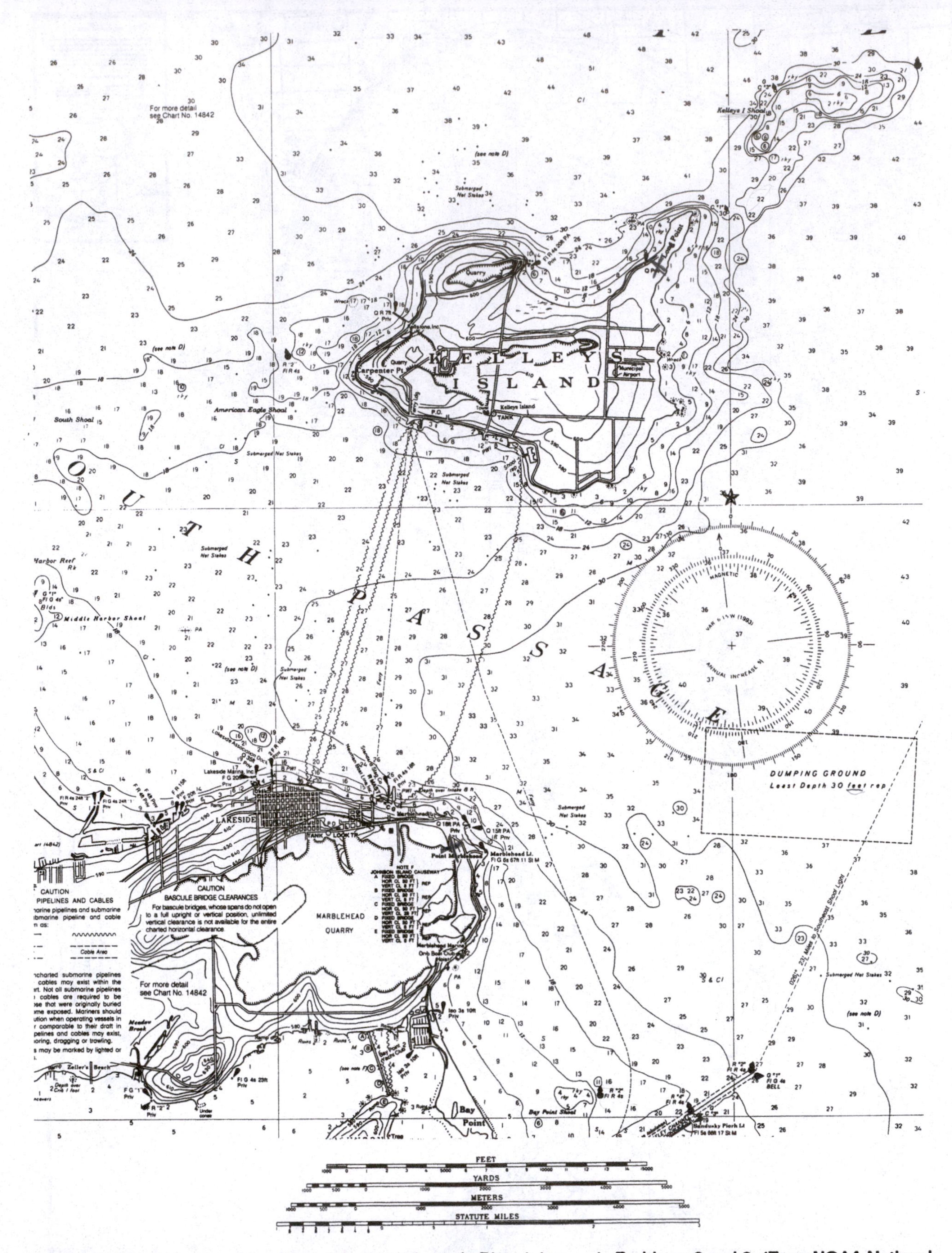

Figure 3.1. Detailed chart of a portion of the area shown in Plate I, for use in Problems 2 and 3. (From NOAA National Ocean Service Hydrographic Chart 14844.)

statute miles per hour ___2.24___

$\frac{3.28}{5280} = 0.0006 \times 60 \times 60$

km/hr ___3.6___

100cm	1m	1km
	100 cm	1000m

1c.) Now, how do you compare the Gulf Stream to your walking speed?

Faster ______ Slower ___✓___ Same ______

1d.) The distance across the North Atlantic, from New England to Europe, is about 3000 nautical miles. If a bottle were to float across the Atlantic in the Gulf Stream, about how many days would it take to cross the ocean?

$\frac{3,000}{1.94} = \frac{1,546.39}{24} = 64.43$

≈ 65days

Using hydrographic charts: A hydrographic chart always has a legend that contains much information that you must read before you can use the chart properly. Refer to the hydrographic chart of Plate I and note the following information in the legend

> **title:** *LAKE ERIE: West End of the Lake Including the Islands*
> **type of projection:** *polyconic*
> **scale:** *1:100,000*
> **sounding unit:** *soundings in feet*
> **navigational information**
> **bar scales** (to show distances)

The small numbers scattered around the chart are called ***soundings***. They show the depth of the water at specific locations.

Problem 2: Answer the questions below using the information in Plate I, and in Figure 3.1, which is a more detailed chart of part of the area shown in Plate I.

12.5 cm

2a.) What is the straight-line distance from the northern tip of Point Marblehead (on the peninsula north of Sandusky) to the northern tip of Long Point on Kelleys Island?

meters ___≈10,500 meters___

2b.) The draft of a ship is the depth from the waterline to the deepest part of the keel. What is the greatest draft (in feet) a ship can have and still be assured of being able to go through the passage between Kelleys Island and the Kelleys Island Shoal? Hint: remember that the water depth at a site between two contours must be shallower than the deeper contour and deeper than the shallower contour, but we cannot determine the depth at the site with greater accuracy than that.

Anything less than 20 feet

A ***compass rose*** shows the directions of ***true North*** and ***magnetic North***. The deviation between them is called the ***magnetic declination***. The compass rose also usually displays information about the annual change in the declination.

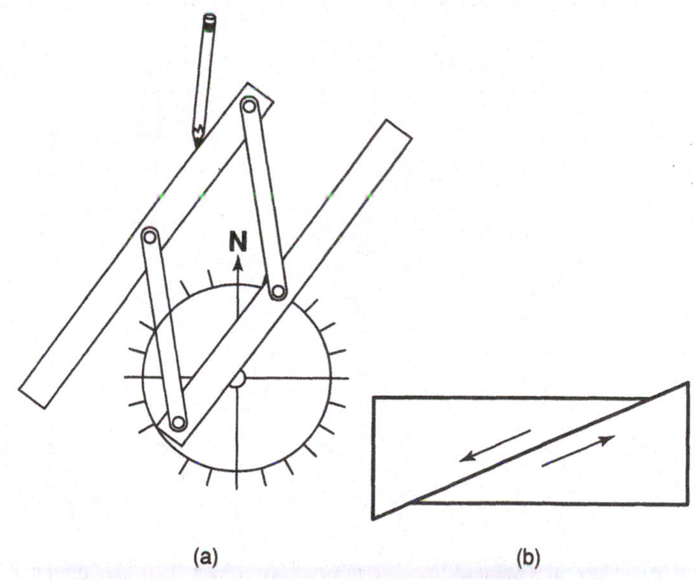

Figure 3.2. A set of parallel rulers (a), or a substitute made from a cardboard rectangle (b), can be used to transfer a compass direction from the compass rose to any other spot on the chart.

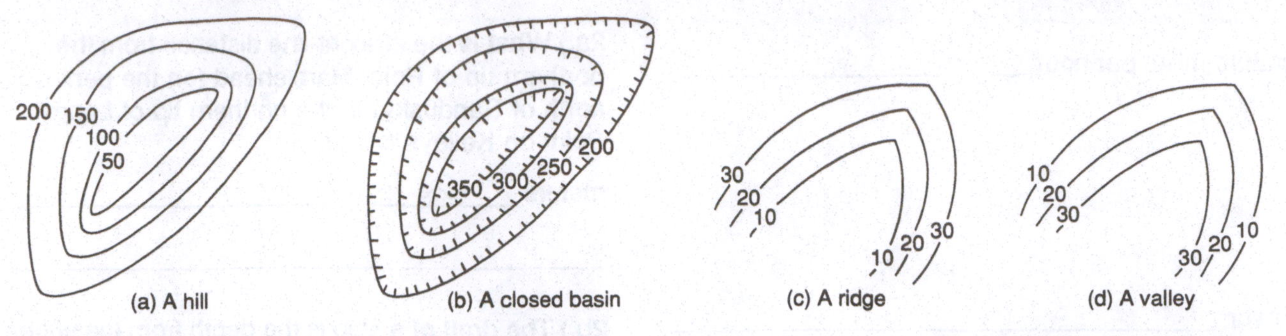

Figure 3.3. Contours illustrating, from left to right, a hill, a closed basin, a ridge and a valley. Numbers are water depths.

The compass rose is very useful for plotting a compass course on a chart. The navigational equipment of most ships, and many small boats, includes a set of parallel rulers. Line up one leg of the ruler along the desired direction on the compass rose as seen in Figure 3.2a. Move the other leg of the ruler so that it passes through the present position of the ship, and use that ruler to draw a line with a pencil. If the ship steers the compass course indicated, it will travel along the line you have drawn. An inexpensive substitute for a set of parallel rulers can be made by cutting a rectangular piece of cardboard along the diagonal between two corners, as shown in Figure 3.2b.

Problem 3:

3a.) On Figure 3.1, plot the course of a ship traveling southeast (true) through the passage between Kelleys Island and Kelleys Island Shoal.

3b.) Based upon the information on the compass rose in Figure 3.1, what is the magnetic declination in that region today? Show your calculations.

____

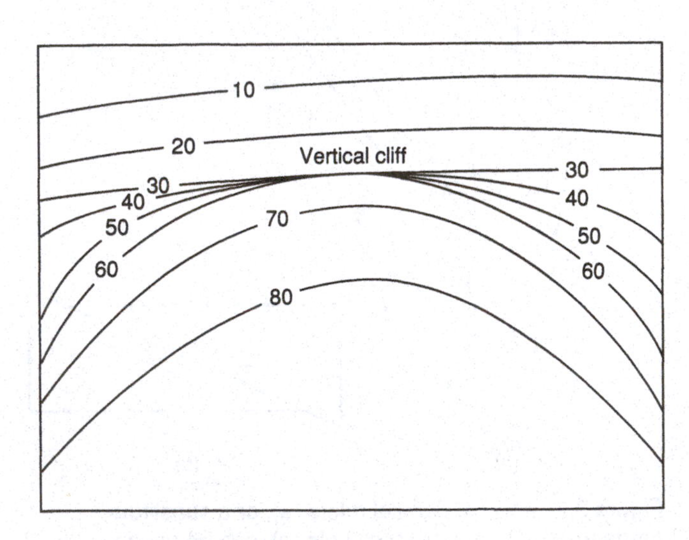

Figure 3.4. Contours may merge, but only to depict a vertical cliff.

____

____

____

Contours and Bathymetry: Bathymetry is the measurement of water depths. One way to represent depths on a two-dimensional surface (i.e., on a piece of paper) is by use of a contour map. Contours are lines drawn connecting points of equal depth. The zero-depth contour is the shoreline. ***Hachures*** (short dashes perpendicular to the contours) indicate that the contours depict elevations within a closed basin rather than around a hill, as shown in Figure 3.3. Contours may merge, but only to depict a vertical cliff (Figure 3.4). However, contours can never cross, for if they did an area of the bottom would be at two different water depths at the same time (Figure 3.5). That, of course, is impossible.

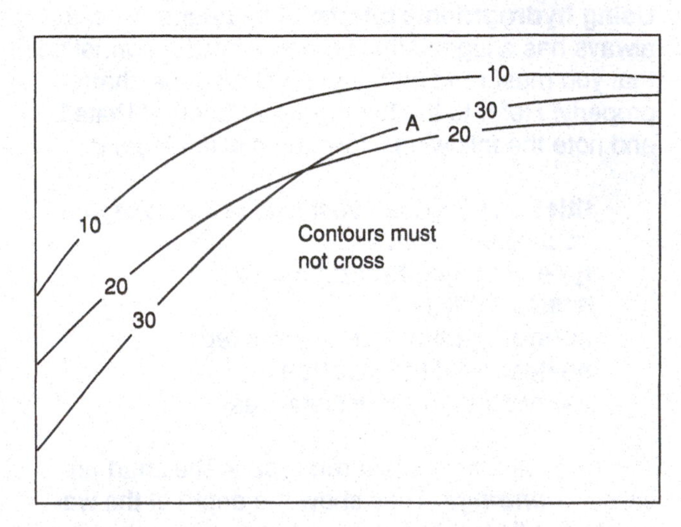

Figure 3.5. Contours may never cross, for that would be logically impossible. For example, in this figure, point A would be at a water depth of 30 meters (using the 30 m contour) and at the same time it would lie between 10 and 20 m as determined from the 10 and 20 m contours. These two interpretations are in conflict with one another.

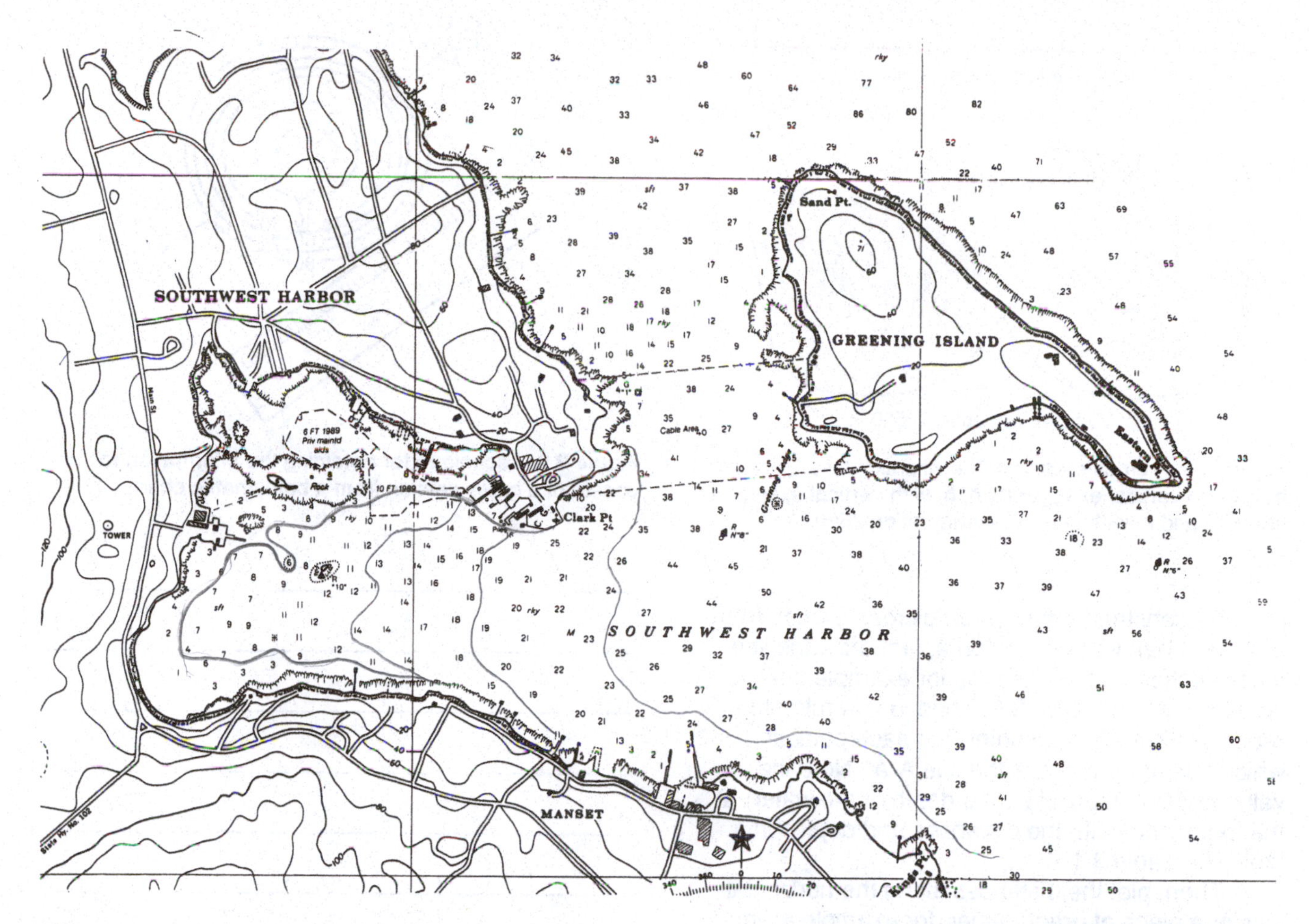

Figure 3.6. Soundings in the vicinity of Mount Desert Island for use in Problem 4. Adapted from National Ocean Service Hydrographic Chart 13321.

Problem 4: Figure 3.6 is a chart of a region near Mount Desert Island on the Maine coast. The numbers in the offshore area of the chart are depth soundings (in feet). Draw contour lines at intervals of 6 feet (i.e. 6, 12, 18, etc.) to show the bathymetry of the ocean bottom in this area. (The contour interval of 6 feet is frequently used in coastal navigational charts. 6 ft = 1 fathom.)

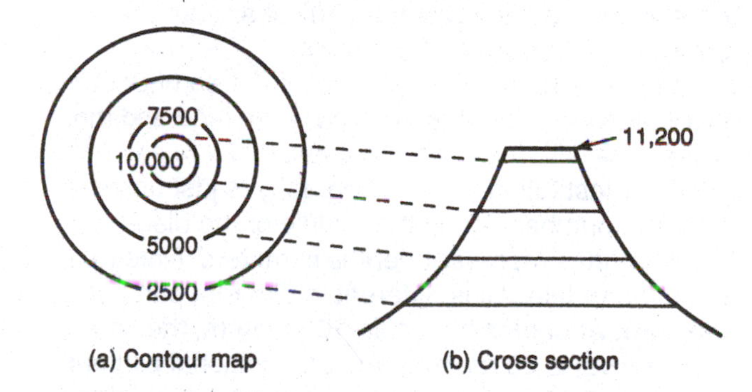

Figure 3.7. Contour map, **a**, and cross section, **b**, of a composite volcano such as Mt. Fuji. Note that contours of features on land show elevations above sea level rather than depths below sea level.

Cross sections and vertical exaggeration: A common way of visualizing bathymetry (or topography) is by means of a cross section. An example is shown in Figure 3.7. A bathymetric cross section is a representation of a vertical slice cut downward through the ocean and the ocean bottom.

In drawing cross sections of geological or oceanographic features, the vertical direction is frequently drawn using a much more expanded scale than the horizontal. The resulting ***vertical exaggeration*** illustrates topographic features that otherwise would appear insignificant or not be visible. Figure 3.8 shows a cross section drawn across the Atlantic Ocean. In Figure 3.8a the vertical and horizontal scales are identical, so there is no vertical exaggeration. In Figure 3.8b the vertical exaggeration is 50 times (x50), which means that 50 miles in the horizontal direction is represented by the same distance as 1 mile in the vertical direction. In Figure 3.8c the vertical exaggeration is x500.

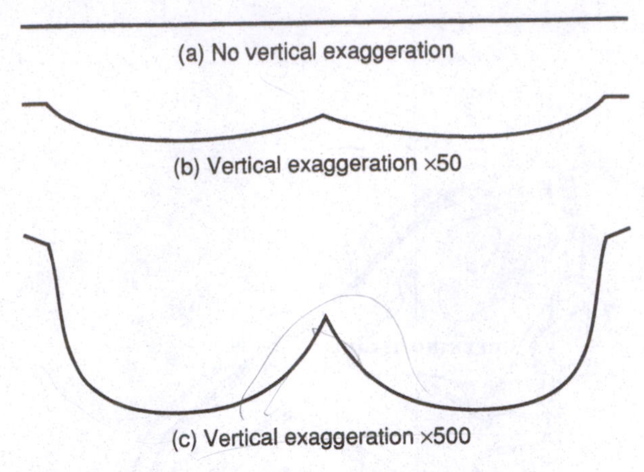

Figure 3.8. A cross section of the Atlantic Ocean, drawn **a.** with no vertical exaggeration, **b.** with vertical exaggeration x50 and **c.** with vertical exaggeration x500.

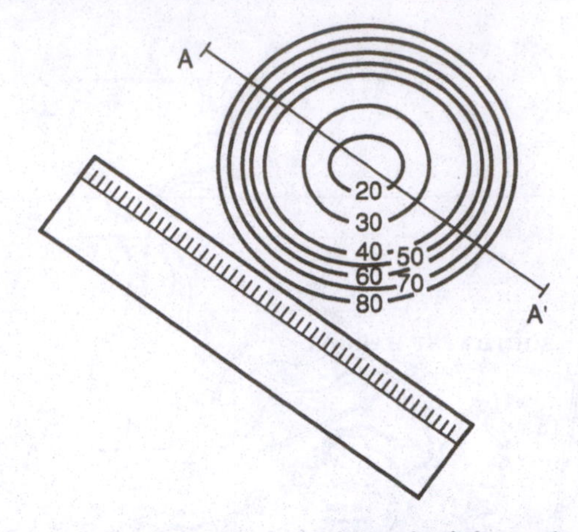

Figure 3.9. Technique for obtaining the information to construct a cross section from a bathymetric chart.

To construct a bathymetric cross section from a chart of bathymetric contours, first draw the line of cross section on the chart, as for example the line labeled A-A′ on Figure 3.9. Then, use a ruler[3] to read the distance from point A to each point at which a contour crosses the line A-A′. Note the value of the contour (i.e., the depth of the water) at that point and write the distances and depths in a table like Table 3.1.

Then, plot the distances along the horizontal axis of a piece of graph paper, for example as in Figure 3.11 (page 27), and the water depths along the vertical axis. Note that when you are plotting a bathymetric cross section you must plot shallower depths toward the top of the vertical axis and greater depths toward the bottom.

Figure 3.10 is a map of a coastal area of Southern California. The shaded area is the land and the white area with the depth contours is the ocean. (Scripps Institution of Oceanography is just onshore from the number "10" in the middle of the diagram.) Contour lines represent depths in meters. Note that the contour interval is different in different parts of the map. At depths less than 100 meters, the contour interval is 20 meters, and an additional contour is shown at 10 meters. At depths greater than 100 meters, the contour interval is 50 meters.

Problem 5: Describe in words the topography depicted by the bathymetric contours in Figure 3.10.

____

____

____

____

____

____

____

____

____

____

Problem 6: Draw a cross section showing the shape of the ocean bottom along line A-A′ in Figure 3.10. The line is 3 km and is marked off in 100 m intervals. Start by tabulating depth as a function of distance in Table 3.1. Then draw the cross section on the graph paper provided in Figure 3.11. Draw a cross section with no vertical exaggeration (upper graph), and with vertical exaggeration of x10 (lower graph).

[3] It is much easier to construct the cross section if your ruler is a decimal ruler, i. e., if it shows divisions in millimeters and centimeters or tenths of an inch.

Table 3.1

Distance from A (m)	*Depth (meters)*	*Distance from A (m)*	*Depth (meters)*
0	10	1,600	19
100	13	1,700	22
200	16	1,800	30
300	21	1,900	38
400	58	2,000	40
500	100	2,100	58
600	100	2,200	100
700	38	2,300	100
800	37	2,400	55
900	37	2,500	35
1,000	30	2,600	30
1,100	23	2,700	22
1,200	20	2,800	18
1,300	19	2,900	16
1,400	18	3,000	14
1,500	18		

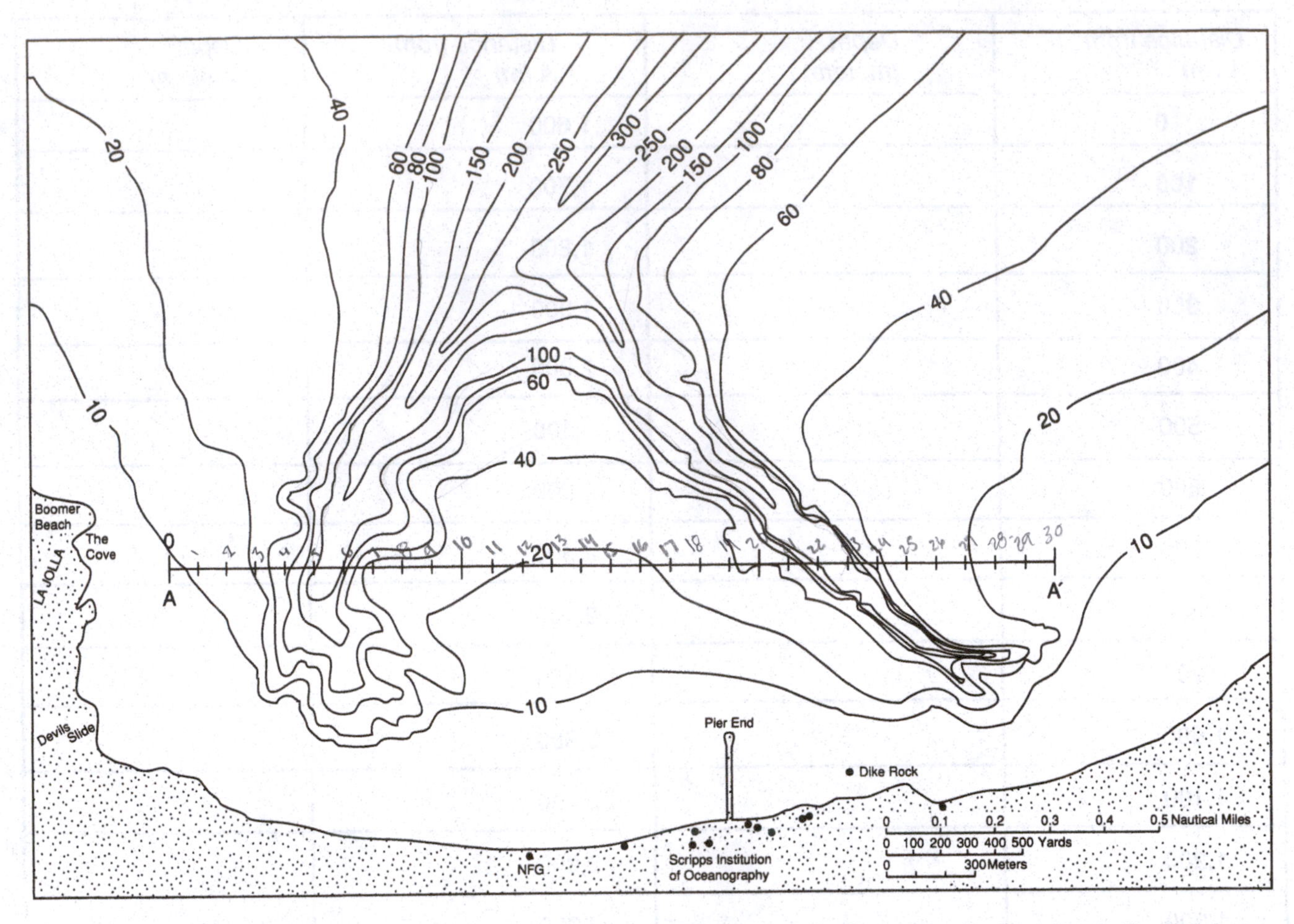

Figure 3.10. Bathymetric chart of the ocean bottom near Scripps Institution of Oceanography (LaJolla, California), for use in Problem 5. Contour interval is 10 m for depths to 20 m, 20 m for depths between 20 m and 100 m, and 50 m for depths greater than 100 m. (Modified from F.P. Shepard and R.F. Dill, *Submarine Canyons and Other Sea Valleys,* Rand McNally, 1966.)

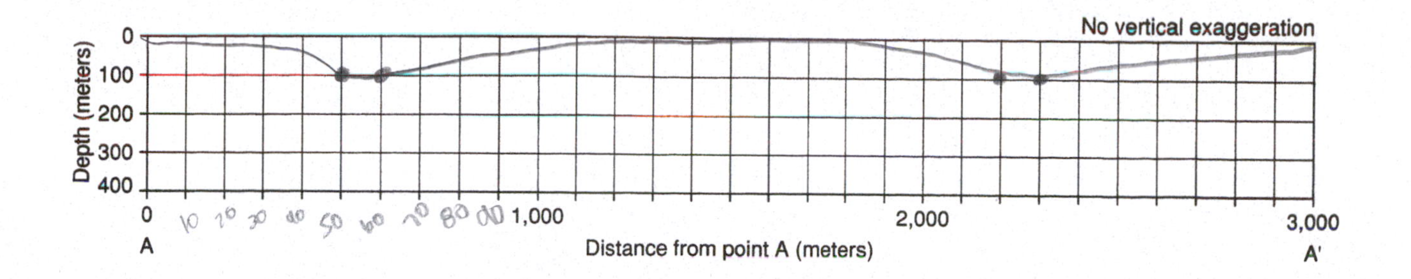

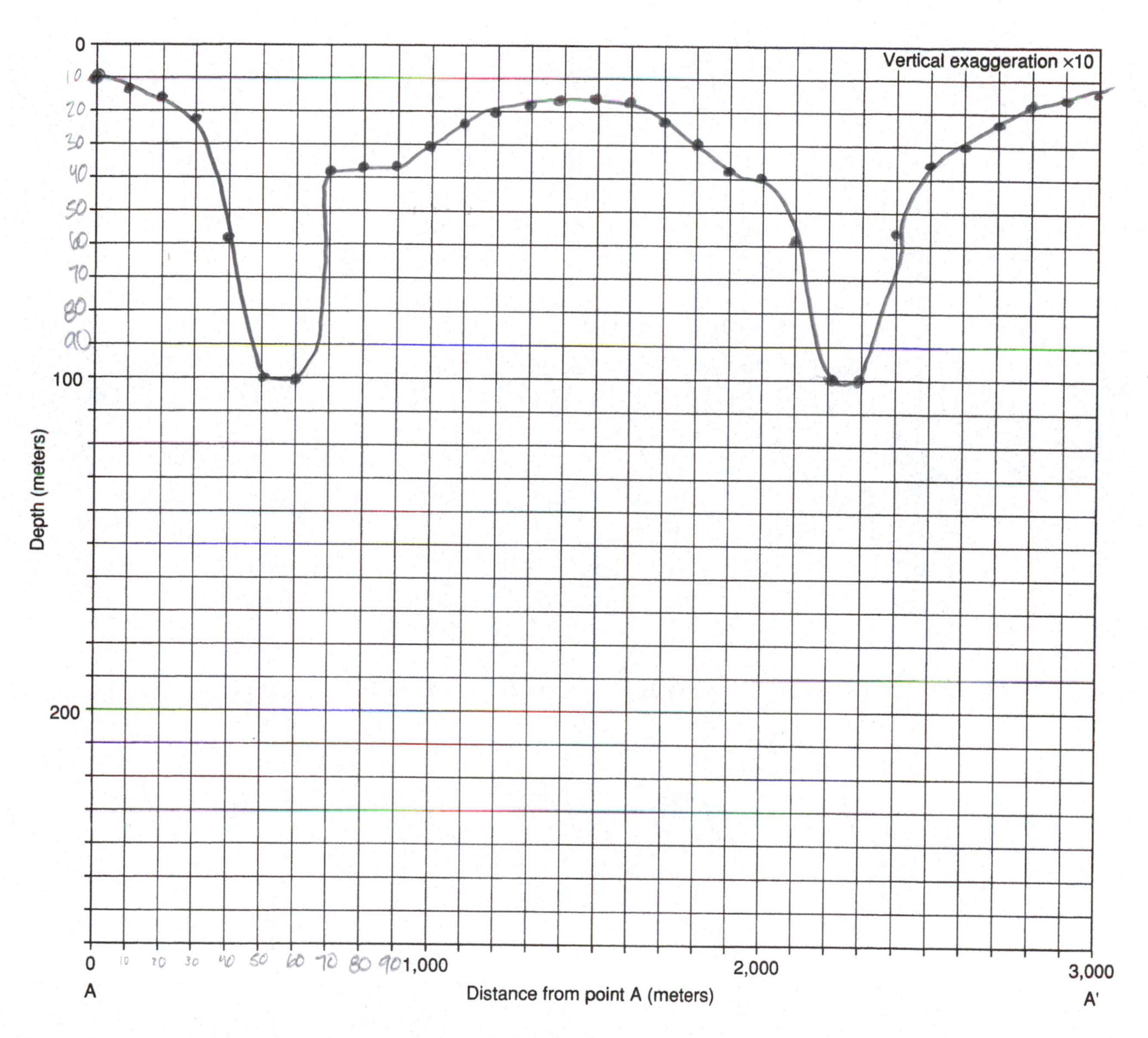

Figure 3.11. Graph paper for plotting cross sections along line A-A′ (Problem 6).

Name: ______________________________ Date: _______________ Section: _____________

Exercise 4
MAGNETICS AND PLATE MOTION RECONSTRUCTIONS

Introduction: At the surface of the Earth, plates of rigid ***lithosphere*** (rock sphere), typically miles in thickness, move horizontally upon a plastic layer of the Earth's ***mantle***, the ***asthenosphere*** (weak sphere). The plates form from volcanic lava along the axes of oceanic mountain ranges, from which they move away in a process called ***seafloor spreading***. The plates typically converge with one another and may be ***subducted*** (move downward) back into the mantle at ***ocean trenches***. These processes are referred to as ***global plate tectonics***, and they are responsible for constructing many of the major features of the Earth, including most mountain ranges and most volcanoes. Earthquakes are commonly one result of the process.

Purpose: In this exercise you will look critically at the kinds of magnetic data used to document the occurrence of the seafloor spreading process. You will then use magnetic data and other geological information to calculate the rate of seafloor spreading in different parts of the world.

The earth's magnetic field: The magnetic field of the earth is often described as ***dipolar*** (i.e., having two poles). In other words, the earth has a magnetic field that resembles the field it would have if a permanent bar magnet (with two poles, North and South) were imbedded near the center, as shown in Figure 4.1. [Note: *The earth does not really have a permanent magnet in it.* The magnetic field is actually generated by motions within the iron-nickel fluid outer core.]

The line through the north and south magnetic poles is the ***axis of the magnetic dipole***. It is not coincident with the earth's rotational axis. It is inclined at present by about 23° from the rotational axis, but that number changes with time. There is evidence, however, that over a period of several thousand years the average direction of the magnetic dipole does coincide with the rotational axis. In other words, even though a compass needle does not point to the North Pole today, if you were sitting with a compass at the same site for several thousand years, the average direction of the compass needle during that time *would be* true north.

Some minerals are weak permanent magnets. The best known of these, and the strongest magnet among the minerals, is the common iron oxide mineral ***magnetite***, also known as lodestone. Pieces suspended from string were used as crude compasses by the Chinese and, beginning in the fourteenth century, in Europe.

Magnetite occurs commonly, although not abundantly, in many kinds of rocks, including basalt, a common volcanic rock. Basalt underlies the sediment that covers most of the deep ocean floor, constitutes the bulk of most mid-ocean islands like the Hawaiian Islands, and is abundant on the continents. When basaltic lava cools, magnetite crystals grow. When these crystals cool below their ***Curie temperature*** (the temperature above which a material loses its ability to be a permanent magnet—about 600°C for magnetite) they acquire ***remanent magnetism***, or in other words, become permanent magnets. The permanent magnetic field of the rock is aligned parallel to that of the earth's magnetic

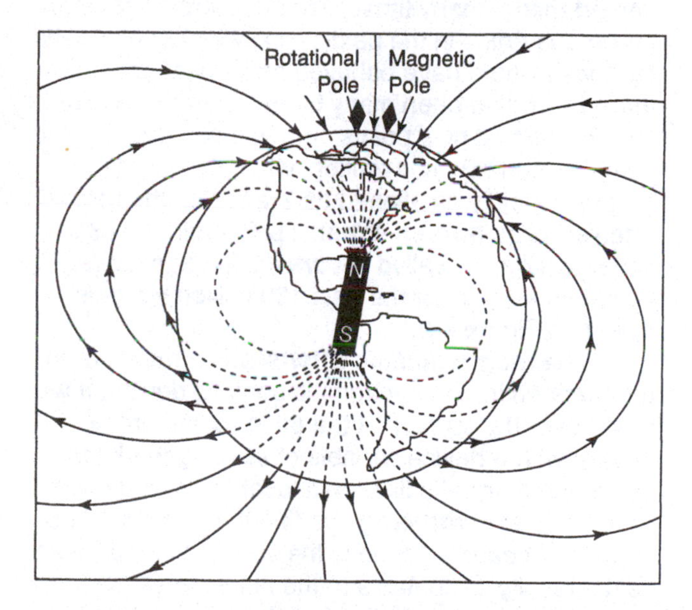

Figure 4.1 The magnetic field of the earth is dipolar. It resembles the field the earth would have if a permanent bar magnet existed in the interior. There is, however, no such permanent magnet within the earth's interior. (Reprinted with the permission of Macmillan Publishing Company from *The Earth,* Fourth Edition, fig. 18.8, by Edward J. Tarbuck and Frederick K. Lutgens. Copyright © 1993 by Macmillan Publishing Company.)

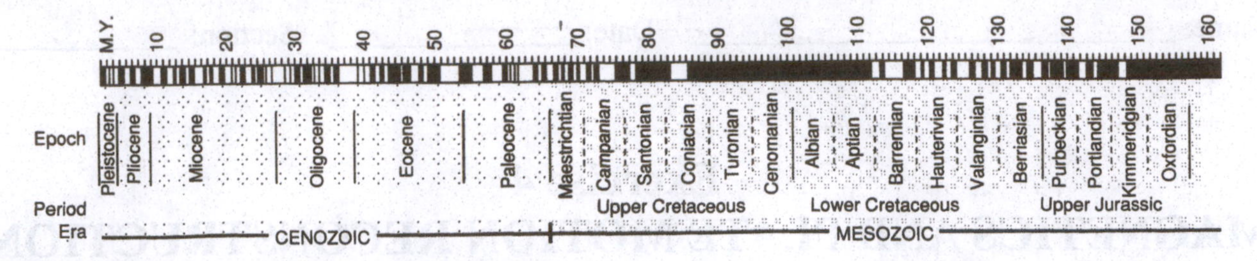

Figure 4.2. Record of intervals of normal and reversed magnetic polarity during the past 160 million years. The black portions of the bar represent intervals during which the earth's magnetic field was normal. The white portions of the bar represent times during which the earth's magnetic field was reversed.

field at the time the rock cools below the Curie temperature. If the rock does not become reheated above the Curie temperature or become grossly altered by weathering (i.e., soil forming processes), it retains the magnetic properties it acquired when it first cooled. Therefore, such rocks act very much like tape recorders, telling us the direction of the earth's magnetic field at the time they cooled.

During the past thirty years, large numbers of measurements have been made of the direction of the remanent magnetism in rocks from all over the world and of all ages. As a result of these measurements it has been possible to demonstrate that the direction of the magnetic field of the earth has reversed many, many times. That is, there have been numerous times in the past during which a magnetic compass would have behaved as it does today, but there have also been many times when the end of the needle that now points north would have pointed south. By convention, when the earth's magnetic field was aligned as it is today, the field is referred to as ***normal***. When it pointed in the opposite direction it is called ***reversed***. A magnetic reversal time scale for the past 160 million years is shown in Figure 4.2.

The ***magnetic field intensity*** is a measure of the force acting on a magnetic sensing device. If we travel over the surface of the earth (continent or ocean) with a device capable of sensing the intensity of the magnetic field on a local scale (a device called a ***magnetometer***), we find that the total field intensity at each location is the sum of the magnetic field intensity attributable to the planet itself (called the ***dipolar magnetic field***) and the field intensity attributable to the remanent magnetism of the rocks present at the location where we make our measurements. (See Figure 4.3.)

We use the term ***anomaly*** to describe a condition in which a measured quantity has a value greater or smaller than an expected value. In places where the rocks were formed at a time when the earth's dipolar field was normally magnetized, the total field is greater than the dipolar field. We say that we measured a ***positive magnetic anomaly***.

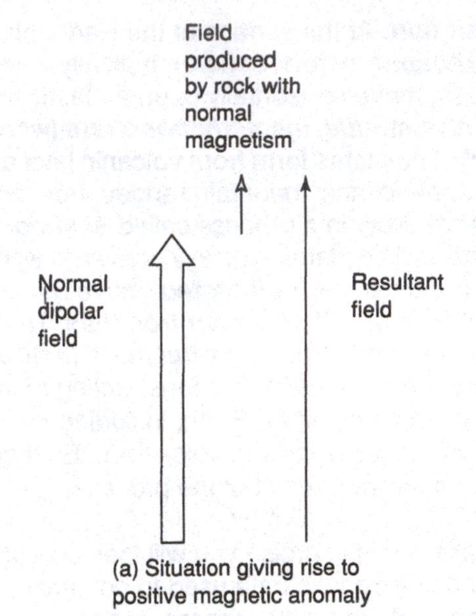

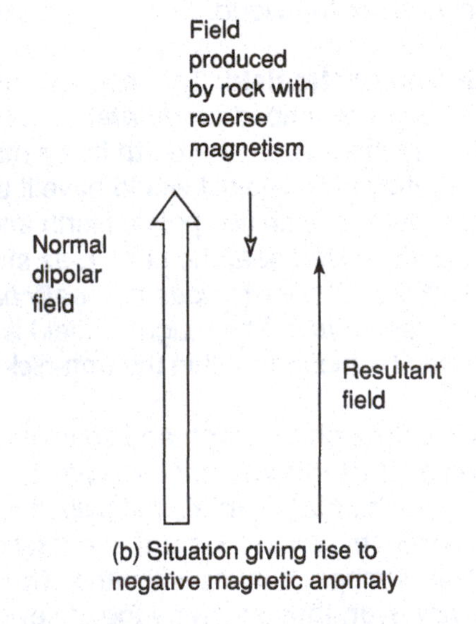

Figure 4.3 Positive and negative anomalies are created as the result of the interaction of the earth's magnetic field and the magnetic field produced locally by the remanent magnetism of the rocks.

Where the rocks were formed at time of reversed magnetization, the direction of the magnetic field of the rocks is opposite to that of the dipolar field and hence partially cancels out the dipolar field. The intensity of the overall magnetic field sensed by our magnetometer is lower than the dipolar field intensity and we therefore say that we measured a ***negative magnetic anomaly***.

Magnetic anomalies on the seafloor are aligned parallel to the axes of oceanic mountain ranges called ridges and rises and are perpendicular to the fracture zones that offset segments of the ridges and rises. Magnetic intensity information is typically displayed in one of two ways (Figure 4.4). The intensity data themselves may be plotted, as in Figure 4.4a (The unit of magnetic field intensity is the *gamma*), or zones of positive and negative magnetic anomalies may be shown by shading (Figure 4.4b). The former shows more detail, which is sometimes useful, but the detail can sometimes obscure important patterns. The second method, showing only the broad patterns of ***magnetic stripes***, is frequently useful for illustrating the principal relationships. Which method is best depends on what you are trying to see. Patterns of magnetic stripes emphasize regional patterns but obscure local details.

Symmetry of magnetic intensity patterns: Symmetry of the magnetic intensity patterns about the spreading axes provides one of the most compelling arguments for the occurrence of seafloor spreading. When the pattern is symmetrical, the part of the pattern to the right of the spreading center can be seen to be the mirror image of the pattern to the left of the spreading center. The easiest way to determine whether a magnetic intensity record has a mirror plane is to compare the complete record with its mirror image. Without bathymetric data to help one locate the axis of seafloor spreading, the search for a center of symmetry in the magnetic record must be done rather subjectively. One selects a point on the magnetic intensity record that appears to be a likely candidate to be a mirror plane. Then, by lining up that possible mirror plane on the actual record with the same point on the mirror image it is easy to see whether a peak-by-peak and valley-by-valley match can be made between the record and its mirror image. If so, one can be confident that the proposed mirror plane is indeed a mirror plane, and therefore marks the location of the spreading axis. An example is shown in Figure 4.5. One can be confident that the point marked by the arrow on the record and its mirror image is a mirror plane because of the close matches of positive and negative anomalies on either side of the ridge.

Problem 1: In this problem you are given magnetic records from two different parts of the ocean. The horizontal axis on each of these records is distance,

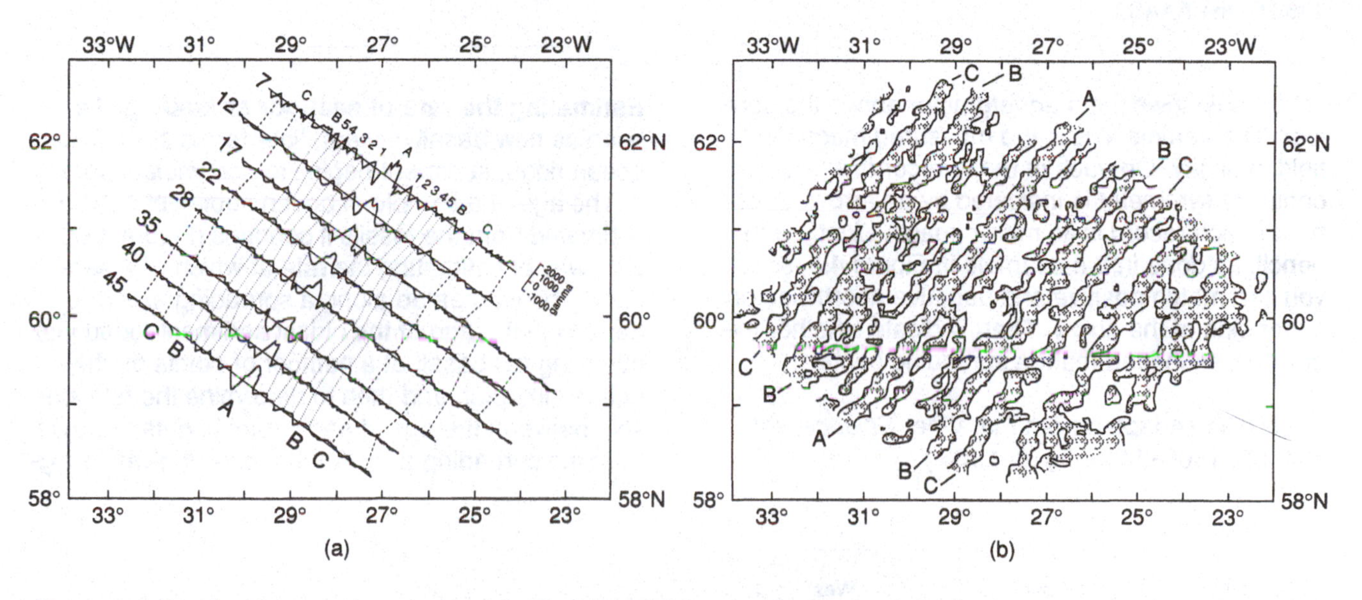

Figure 4.4. Results of magnetic surveys of the Reykjanes Ridge, south of Iceland. On the left are the records of magnetic intensities plotted along several ship tracks running across the ridge axis and perpendicular to its crest. On the right the data have been converted to magnetic anomaly patterns (or magnetic stripes). Stippled bands show areas in which the measured magnetic field intensity is greater than the regional average. Light bands show areas where the measured magnetic field intensity is less than the regional average. The ridge crest lies along a line connecting the points labeled A in each diagram. The positive magnetic anomalies defined by lines connecting points labeled B and labeled C in each of the diagrams are symmetrical about the ridge crest. (Reprinted from *Deep Sea Research*, Volume 13, J.R. Heirtzler, X. lePichon and J.G. Baron, Magnetic anomalies over the Reykjanes Ridge, Pages 427–443, Copyright 1966, with kind permission from Elsevier Science Ltd., The Boulevard, Langford Lane, Kidlington 0X5 1GB, UK.)

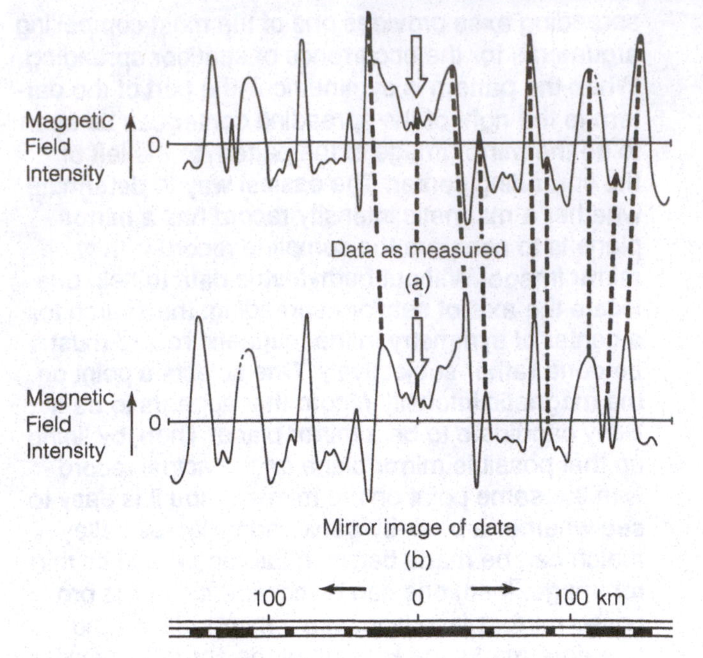

Figure 4.5 Magnetic field intensity along a ship track running perpendicular to the Mid Atlantic Ridge. The lower record is simply the mirror image of the upper record. The vertical arrow near the center of each record marks the center of symmetry. Note how, in this record, each of the magnetic fluctuations to the left of the vertical arrow corresponds to a similar fluctuation to the right of the arrow. This demonstrates that the magnetic anomalies are symmetrical about the ridge axis. (Reprinted with permission from *Science*, Volume 154, F.J. Vine, Spreading of the ocean floor: new evidence, Pages 1405–1415, Copyright 1966 by the AAAS.)

as the ship tows the magnetometer above the seafloor. The vertical axis is the measured magnetic field intensity. On each of these records a possible center of symmetry is indicated by an arrow. Underneath each record is its mirror image. Lightly with pencil, as done in the example in Figure 4.5, see if you can match positive and negative anomalies on either side of the arrow. Then, indicate whether the arrow does indeed mark a spreading center.

Record #1 (Adapted from F.J. Vine, *Science*, vol. 154, pp. 1405–1415, fig. 8, 1966.)

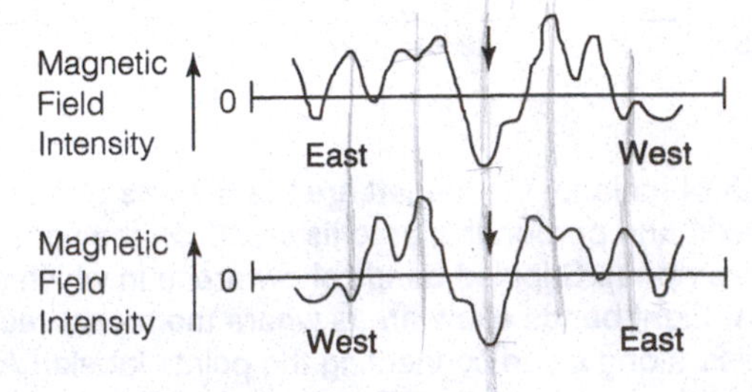

Does the arrow lie at a spreading center? Yes ✓
No______
Explain in a sentence or two why you have given the answer that you did.

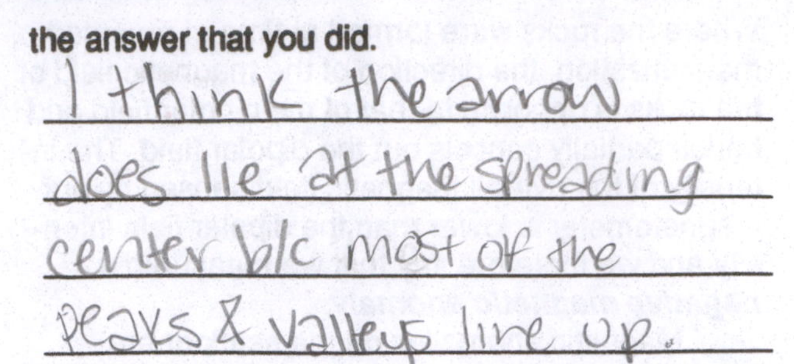

Record #2 (Adapted from F.J. Vine, *Science*, vol. 154, pp. 1405–1415, fig. 17, 1966.)

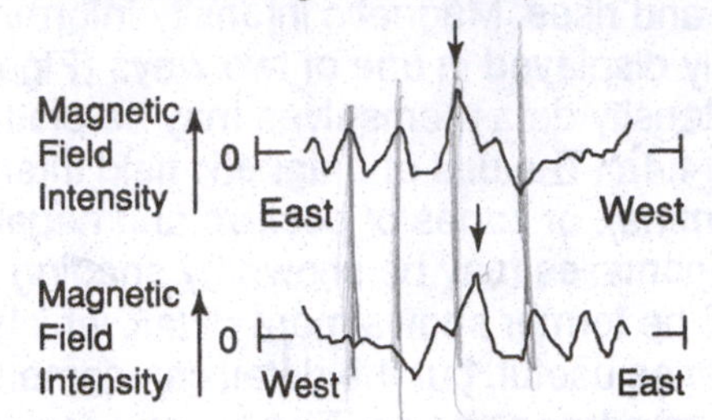

Does the arrow lie at a spreading center? Yes______
No ✓
Explain in a sentence or two why you have given the answer that you did.

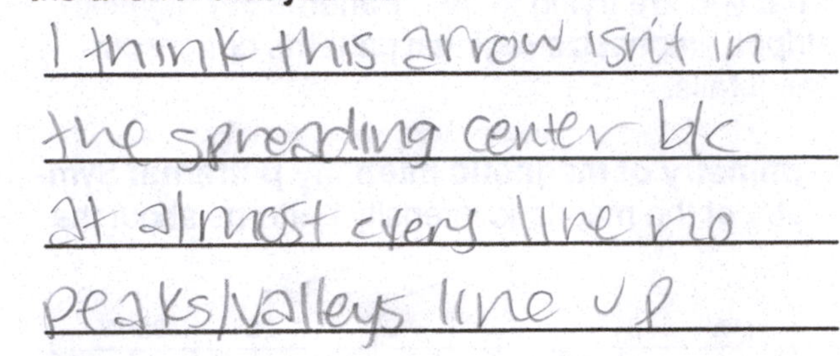

Estimating the rate of seafloor spreading: As soon as new basaltic ocean floor forms at the mid-ocean ridge, sediment begins to accumulate upon it. The age of a sample of ocean sediment can be estimated from the fossils it contains (Figure 4.6). One way of estimating the rate at which new seafloor is being created along a spreading axis is to determine the age of the oldest sediment found just overlying the basalt at a number of points on the plate of interest, and then to determine the relationship between the age at each point and its distance from the spreading axis. An example is given in Fig-

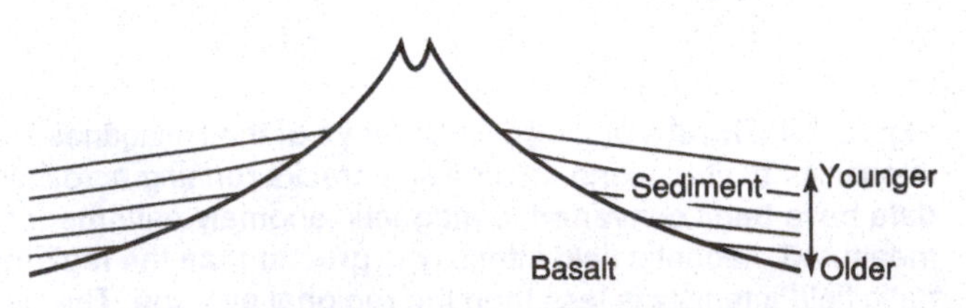

Figure 4.6. Sediment begins to accumulate on the basaltic lithosphere after it forms at the mid-ocean ridge. So the age of the oldest sediment at any location is approximately the age of the basaltic lithosphere.

ure 4.7 for the South Atlantic. (Note that the rate at which a plate is growing is called the ***half-spreading rate***. The ***total spreading rate*** is twice that value because both plates which meet at the spreading axis are growing at approximately the same rate.

Problem 2: In your own words, explain why the half-spreading rate can be determined using measurements at several sites of the age of the oldest sediment and the distance of the site from the spreading axis.

Scientists can use math to calculate how long it's been since sediment has been released by taking several samples equally distanced apart and determining a rate $\frac{y_2 - y_1}{x_2 - x_1}$ = slope rate

Problem 3: Calculate the half-spreading rate of the South Atlantic from the data shown in Figure 4.7. Show your work.

$$\frac{50-10}{1{,}000-200} = \frac{40}{800} = 0.05 \frac{km}{myr}$$

$$\frac{\text{distance}}{\text{time}} \quad \frac{1000-200}{50-10} = \frac{800}{40} = 20 \frac{km}{myr}$$

Problem 4: As you do this problem, assume that the South Atlantic has had the same half-spreading rate from the time of the initial opening of the rift between South America and Africa until today. In order to calculate the length of time it took from the original rifting until the South Atlantic reached its present width, you need one additional (geographical) measurement. What must you measure?

The distance between each continent (SA & Africa)

$$\frac{\text{distance}}{\text{time}}$$

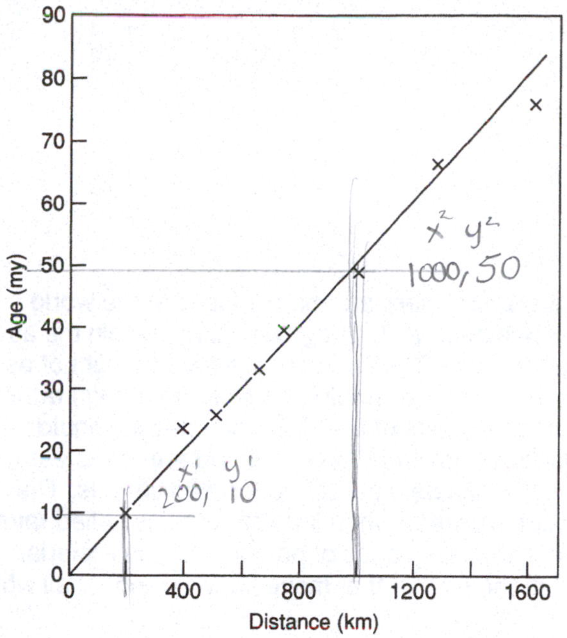

Age (my) — 0, 10, 20, 30, 40, 50, 60, 70, 80, 90

Distance (km) — 0, 400, 800, 1200, 1600

x^1 y^1 200, 10

x^2 y^2 1000, 50

Figure 4.7. Results of drilling several holes through the sediments of the south Atlantic Ocean. The age of sediment resting directly upon the basaltic ocean crust at each of several sites is plotted as a function of the distance from the ridge axis to the drilling site. (From A.E. Maxwell et al., Initial Reports of the Deep Sea Drilling Project, vol. 3, fig. 13.8, 1970).

or how wide the South Atlantic ocean is

You must find a value for that measurement. Write that value here along with the source (e.g., globe, atlas, textbook, etc.) from which you obtained it.

2,850 km for everything

1,425 km for half

The teacher or google

Now, using that measurement and your answer to Problem 3, calculate how long ago the South Atlantic began to form. Show your work.

$$\frac{1{,}425 \text{ km}}{20 \frac{km}{my}} = 71.25 \text{ million years}$$

Hot spots: There are many places in the world where molten rock (***magma***) occurs within the asthenosphere. These zones are best thought of as mixtures of liquid (melt) and solid (minerals) rather than as pockets of liquid. Sometimes the liquid is squeezed upward toward the surface of the earth where it can be extruded to form volcanoes. The molten material, when it is extruded, is called ***lava***. ***Hot spots*** are zones of partially molten material within the mantle that move slowly or not at all while the overlying lithosphere moves over them. As seen in Figure 4.8, a chain of volcanoes may mark the path of the lithosphere above the asthenospheric hot spot. The Hawaiian Islands are part of such a chain. The entire chain is called the ***Hawaiian-Emperor Seamount Chain***. Most of the volcanoes of the chain are submarine volcanoes or seamounts. A few are low-lying islands that barely protrude above the ocean surface. The southeasternmost volcanoes of the chain are the populated islands of the state of Hawaii. The highest point in the state, Mauna Loa, stands about 4,200 meters (13,680 feet) above sea level and about 9.2 kilometers (5.7 miles) above the floor of the surrounding ocean.

A perspective view of the Hawaiian-Emperor Seamount Chain is shown in Figure 4.8. A map of the southeasternmost Hawaiian Islands, with the ages of their volcanic rocks, is shown in the inset of that figure.

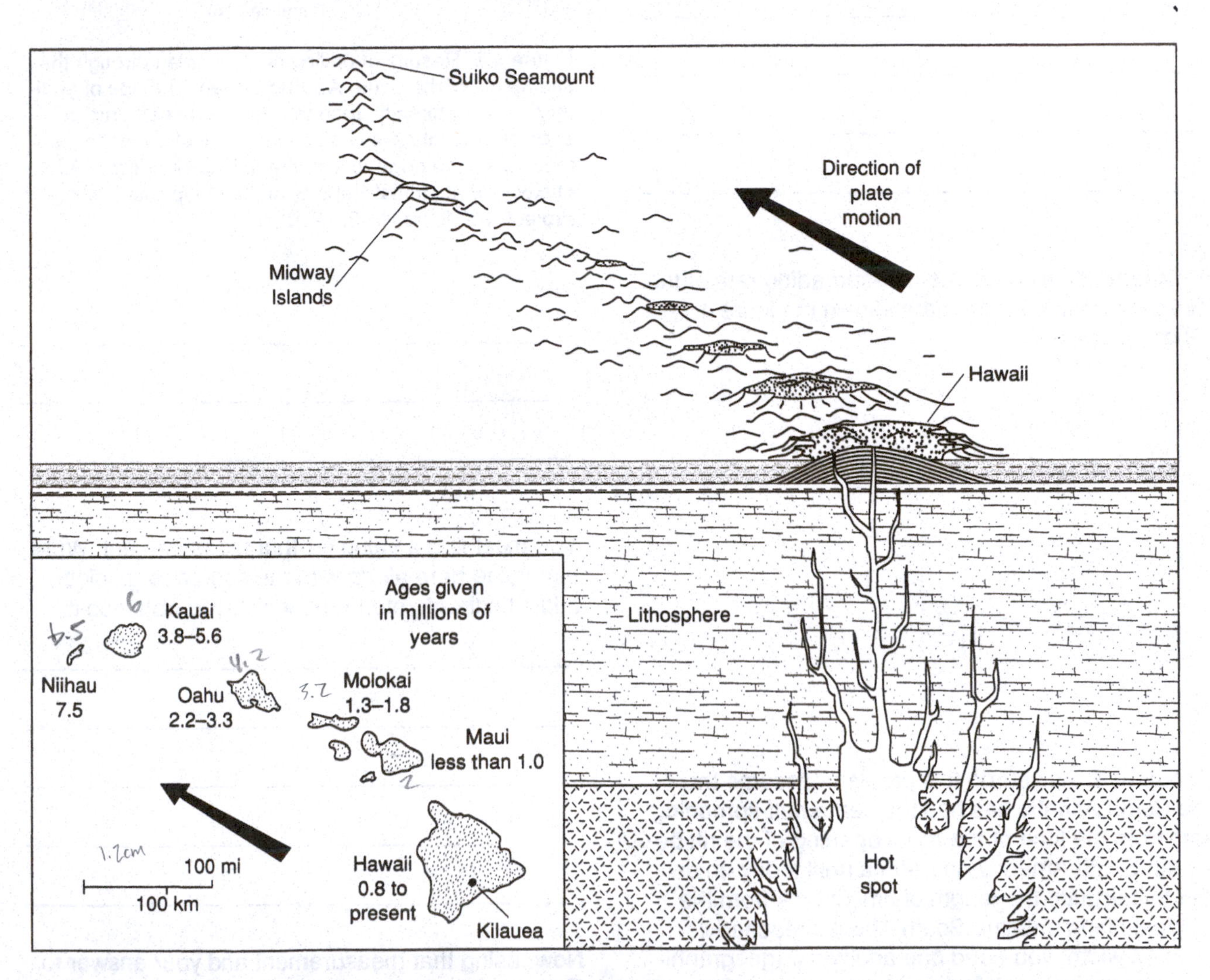

Figure 4.8. The chain of islands and seamounts that extends from Hawaii to the Aleutian trench results from the movement of the Pacific Plate over an apparently stationary hot spot. Radiometric ages of lavas are indicated on the diagram. Volcanism is still very active at Kilauea on the island of Hawaii. (Reprinted with the permission of Macmillan Publishing Company from *The Earth,* Fourth Edition, fig. 18.28, by Edward J. Tarbuck and Frederick K. Lutgens. Copyright © 1993 by Macmillan Publishing Company.)

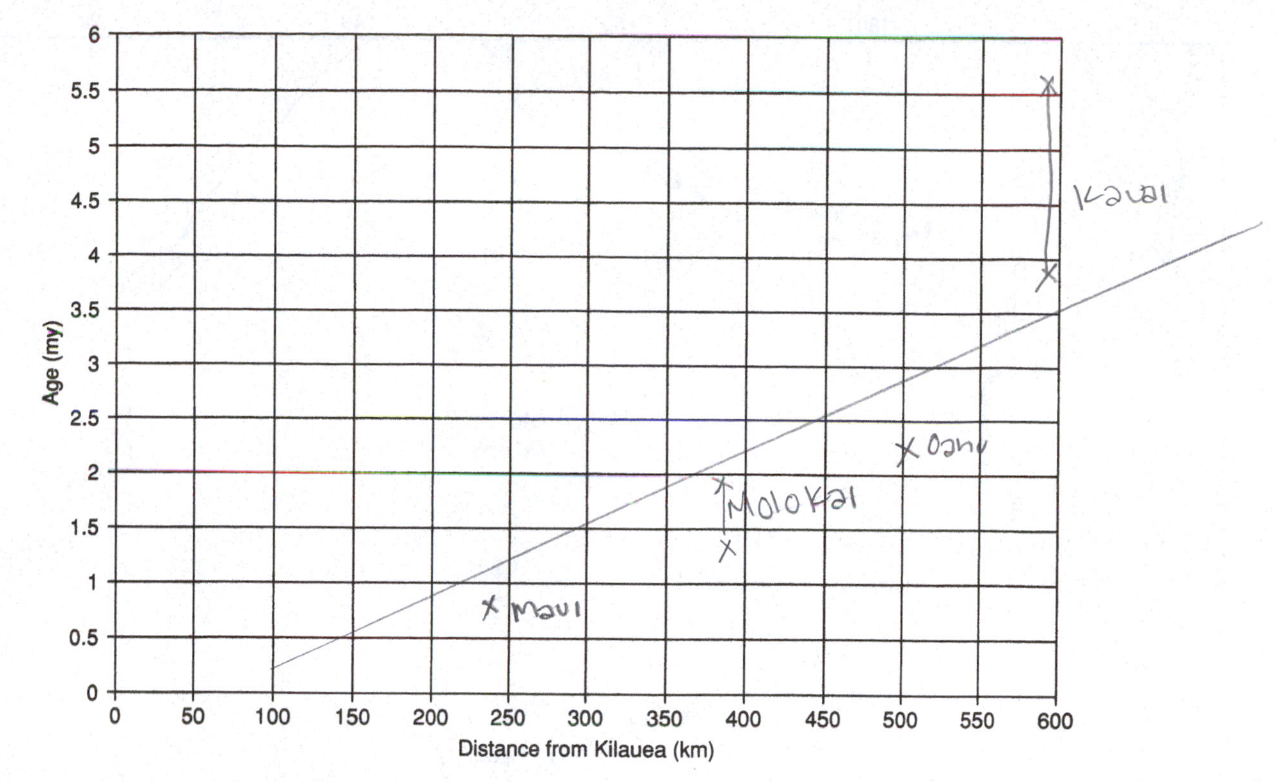

Figure 4.9. Draw your graph for Problem 5 here.

Problem 5: In this problem you will again consider the rate of movement of a plate of oceanic lithosphere. The approach is somewhat different from the one you used in Problem 4. In this problem, assume that all of the islands and the seamounts of the Hawaiian-Emperor Seamount chain were formed as the result of the Pacific plate moving *at a constant speed* over a stationary hot spot.

5a.) On the graph paper of Figure 4.9, plot the age of each of the Hawaiian Islands as a function of the distance from the southeastern end of the southeasternmost island. Each point should be drawn as a cross, the height of which indicates the range of ages of lavas of the island and the width of which indicates the range of distances from some arbitrary starting point you choose. Draw a straight line through the points you have plotted (a visual best-fit line). Using the graph and the line, determine the speed at which the Pacific Plate is moving relative to the hot spot. Show your calculations and results.

In what direction is the Pacific Plate moving?

North West

5b.) It is clear from the bend in the seamount chain that the direction of plate motion changed sometime in the past. Using your answer in 5a, the assumption of a constant rate of plate motion, and the map in Figure 4.10, calculate when the direction of motion changed.

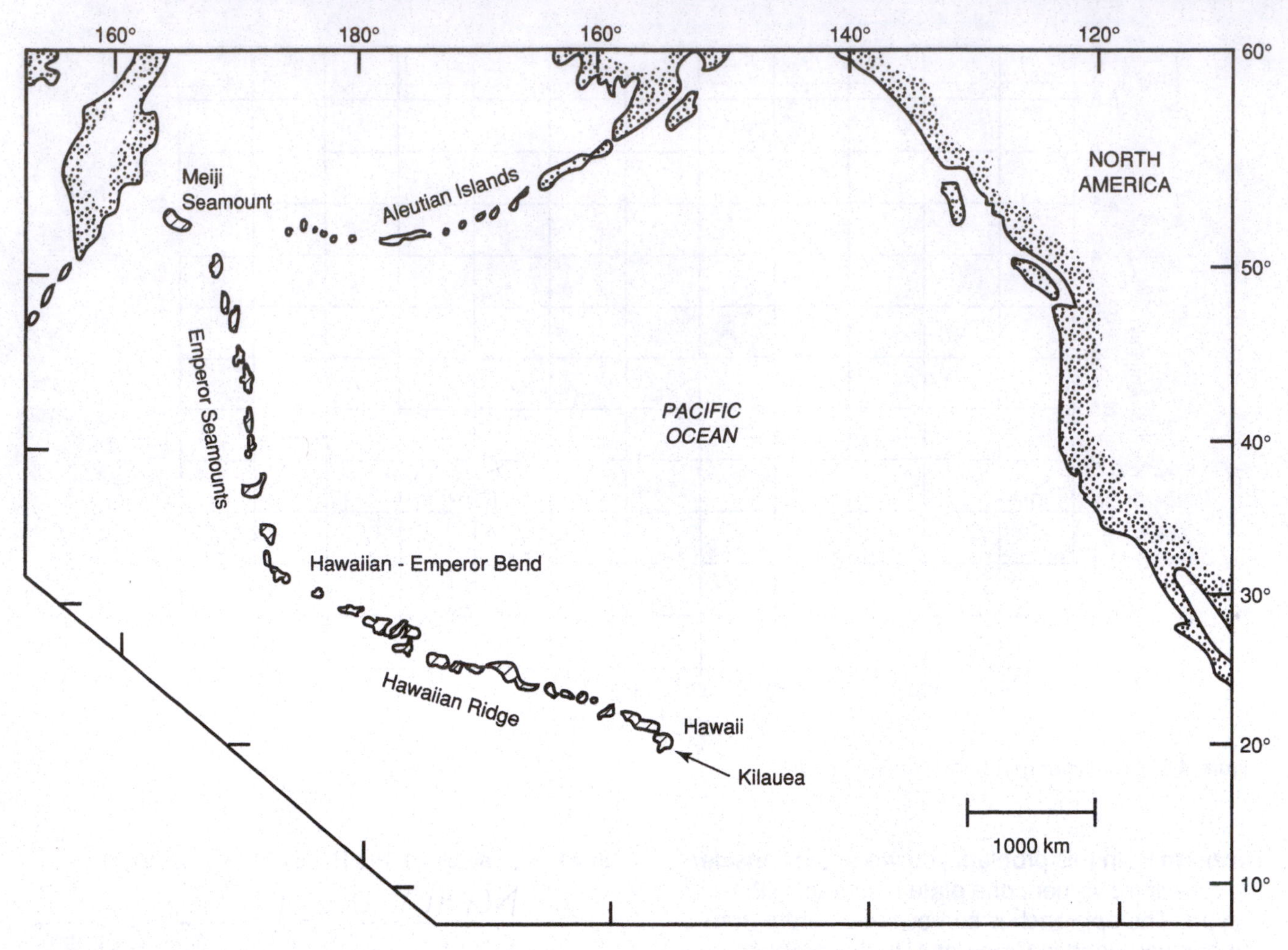

Figure 4.10. Map of the Hawaiian-Emperor Seamount Chain for use in solution of Problem 5b. (Base map from D.A. Clague, G.B. Dalrymple, and R. Moberly, *Geological Society of America Bulletin*, vol. 86, 991–998, fig. 1, 1975.)

Name: ______________________________ Date: ______________ Section: ____________

Exercise 5
SUBMARINE TOPOGRAPHY, ISOSTASY, SEISMICITY, AND PLATES

Purpose: This is the second of two exercises dealing with aspects of plate tectonics. (The first is Exercise 4.) In this exercise you will develop a feeling for the relations between submarine topography, isostasy, and tectonic plate boundaries. You will also relate these tectonic plate boundaries and topographic features to the seismicity (or distribution of earthquakes) of the oceanic lithosphere.

Equipment needed: You will need a protractor and a straightedge.

Features of the seafloor: Plate II is a physiographic chart of the seafloor. A glance shows that the seafloor is far from featureless. Furthermore, the distribution of physiographic features on the seafloor is far from random. The main features of the seafloor are:

Continental Margin, consisting of continental shelf, continental slope, and continental rise. (See Figure 5.1.)

Continental Shelf—These are submerged extensions of the continents. They may vary in width up to hundreds of kilometers, with average widths of about 75 km. The edge of the continental shelf may be defined as the point where the water reaches some arbitrary depth (say 150 meters or 100 fathoms) but is better defined as the point at which there is a definite break in slope—the continental shelf break—which occurs at an average depth of 130 meters.

Continental Slope—These are the primary slopes that extend downward from the continents to the deep seafloor. If there were no water in the ocean and we were viewing the planet from outer space, the continental slopes would probably be the feature of the ocean basins that would stand out most obviously. Slopes average about 4°, and range from 1 to 10°. The continental slopes are frequently cut by ***submarine canyons.***

Continental Rise—These are gradually sloping piles of sediment which have accumulated at the base of the continental slopes. Physiographically they are transitional between the continental slopes and the deep seafloor. The continental rise is typical of ***passive continental margins***, i.e., those at which subduction is not occurring. It is less commonly found at ***active margins***.

Oceanic Ridges/Rises—These are the long, sometimes sinuous chains of mountains which are conspicuous features of all the ocean basins. For example, on Plate II, follow the Mid-Atlantic Ridge down the center of the Atlantic Ocean and around the Cape of Good Hope where it runs northeastward through the Indian Ocean to a point south of India where it branches. From that point, one branch runs northward toward the Gulf of Oman and the

Figure 5.1. Physiographic diagram of the Atlantic coast of North America, from Labrador to Florida. (From *World Ocean Floor* by Bruce C. Heezen and Marie Tharp, copyright 1977 by Marie Tharp. Reproduced by permission of Marie Tharp, 1 Washington Ave., South Nyack, NY 10960.)

Plate II. Physiographic chart of the oceans. (*World Ocean Floor* by Bruce C. Heezen and Marie Tharp, copyright 1977 by Marie Tharp. Reproduced by permission of Marie Tharp, 1 Washington Ave., South Nyack, NY 10960.)

Figure 5.2 Physiographic diagram of the north Pacific Ocean showing the Hawaiian–Emperor Seamount Chain and the Aleutian Trench. (From *World Ocean Floor* by Bruce C. Heezen and Marie Tharp, copyright 1977 by Marie Tharp. Reproduced by permission of Marie Tharp, 1 Washington Ave., South Nyack, NY 10960.)

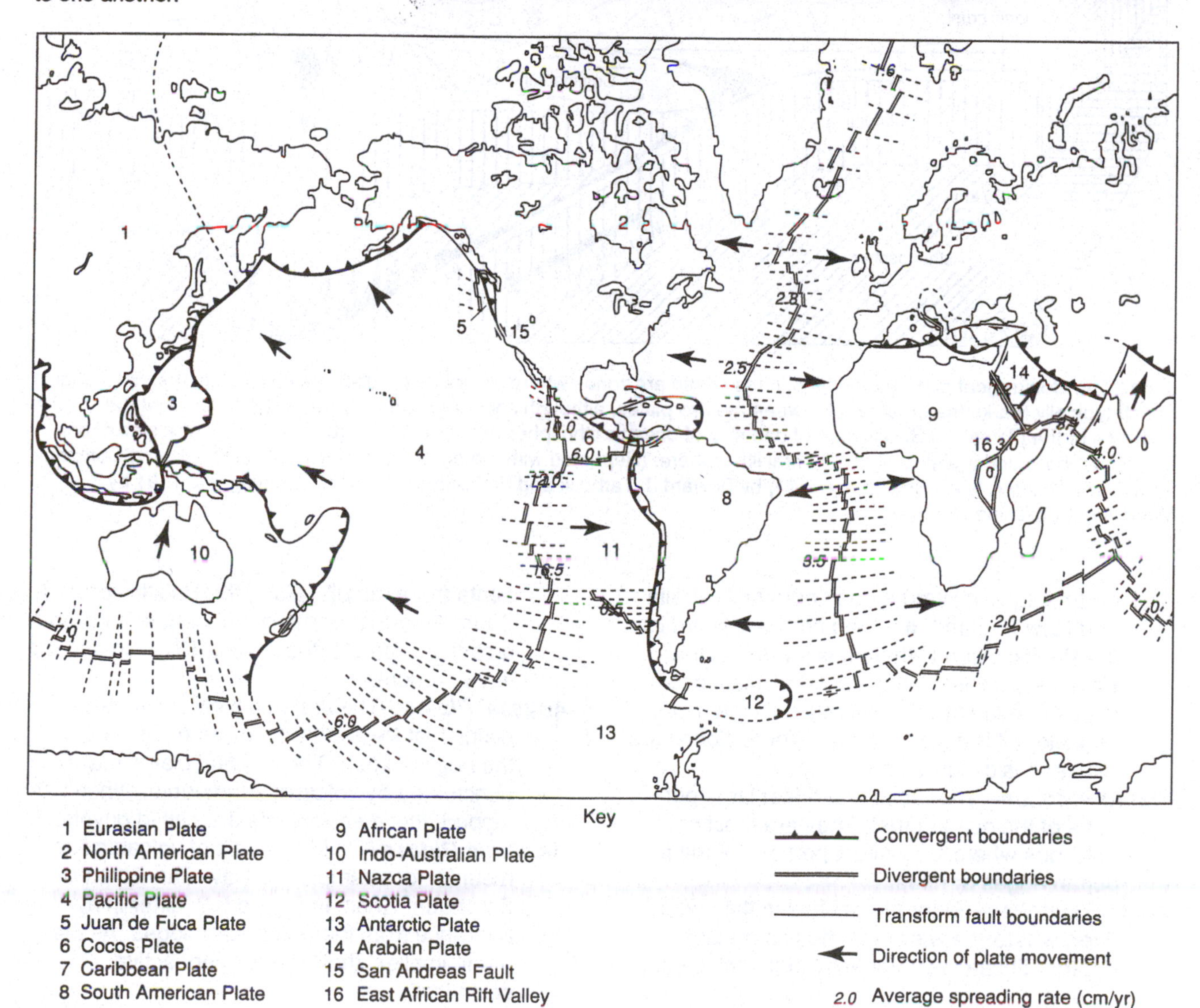

Figure 5.3. The lithosphere can be divided into 14 rigid plates (and a number of small microplates) that move relative to one another.

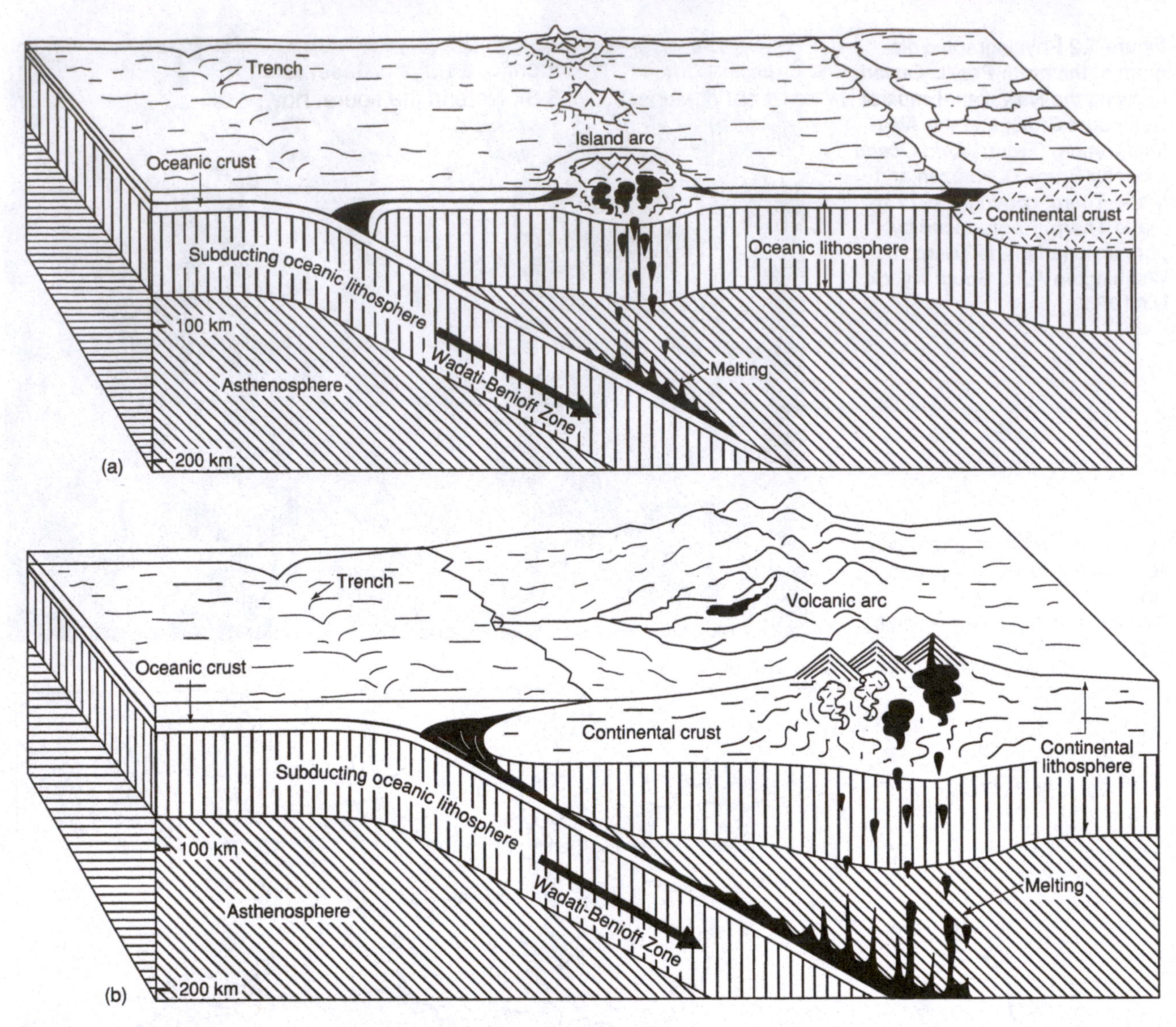

Figure 5.4. Convergent plate boundaries in the ocean are zones where one plate is subducted under another. An oceanic trench typically marks the boundary between the two plates, although some trenches are buried by debris that has eroded from one of the plates. (a) Subduction of one slab of oceanic lithosphere under another. (b) Subduction of a slab of oceanic lithosphere under a slab of continental lithosphere. (Reprinted with the permission of Macmillan Publishing Company from *Earth Science,* Sixth Edition, fig. 6.11, by Edward J. Tarbuck and Frederick K. Lutgens. Copyright © 1991 by Macmillan Publishing Company.)

other southeastward to the south of Australia and New Zealand, the then northwestward into the Pacific Ocean where it is known as the East Pacific Rise. Where the mountain chain is steep-sided (and frequently cut by a medial rift valley) it is called a ridge. Where slopes are gentler it is called a rise.

Fracture Zones—The faults that offset large portions of the ocean crust. They are most conspicuous where they offset portions of the mid-ocean ridges or rises.

Ocean Trenches—Elongate gashes in the ocean floor, where the water reaches its greatest depths. These are frequently adjacent to continents (for example, along the Pacific coast of South America) or the convex edges of island arcs (for example, the Aleutian Trench and the Java Trench).

Abyssal Plains—The flat, relatively featureless plains that frequently occur on either side of the ridge system. The abyssal plains may be punctuated by volcanic seamounts. When flat-topped, these seamounts are called guyots.

Aseismic Ridges—These chains of volcanic seamounts, for example the Hawaiian-Emperor Seamount Chain in Figure 5.2, differ in appearance from the mid-ocean ridges. As the name implies, they are not very active seismically.

Plate margins: The present Earth's lithosphere can be divided into 14 rigid plates (and a larger number of very small microplates that we need not consider) that have moved, and continue to move, relative to one another. (At various times in the geologic past there have been more or fewer lithospheric plates.) The boundaries of the plates that exist on the present Earth are shown in Figure 5.3.

If the plates are moving relative to one another, we can consider the boundaries between them to be of three kinds:

> ***Convergent boundaries*** along which two plates move toward one another.
>
> ***Divergent boundaries*** along which two plates move apart from one another.
>
> ***Transform boundaries*** along which two plates move sideways relative to one another.

Each of the above three types of boundaries has a physiographic expression that can permit you to recognize it.

Convergent boundaries in the ocean are characterized by oceanic trenches, sometimes associated with island arcs. These mark the scars where one plate collides with, and is pushed down under (or subducted under), the other (Figure 5.4). When continental portions of two plates converge, normally neither plate is pushed down under the other. Both plates buckle, and a mountain range results. The Himalayas are a good example.

Divergent boundaries are characterized by mid-ocean ridges or rises. These are the seams along which new oceanic lithosphere is created—the sites of seafloor spreading (Figure 5.5).

Transform boundaries are marked by fracture zones—faults along which one plate slides relative to another. They are most obvious where one segment of a ridge is offset relative to another (Figure 5.6). Note in the figure, however, that only the portion of the fracture zone between two ridge segments is a plate boundary. The other parts of the fracture zone lie entirely in one plate or the other.

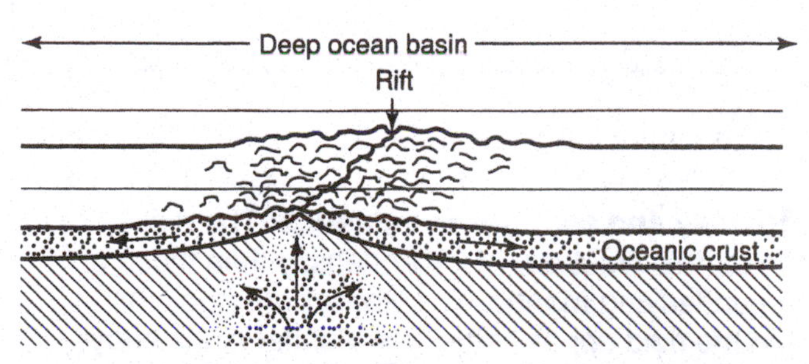

Figure 5.5. Divergent boundaries are sites of creation of new oceanic lithosphere. They are crests of mid-ocean ridges and rises. (Adapted with the permission of Macmillan Publishing Company from *Earth Science,* Sixth Edition, fig. 6.9, by Edward J. Tarbuck and Frederick K. Lutgens. Copyright © 1991 by Macmillan Publishing Company.)

Problem 1: You have already learned that the Atlantic is getting bigger and the Pacific is getting smaller. In this question you are asked to forget that information and to see what you can say about changing sizes of those two ocean basins *simply from observing the nature of the boundaries in the oceans.* That is, solely on the basis of the evidence in Plate II, and Figures 5.3 through 5.6, answer the questions below.

a.) Is the Atlantic Ocean getting:
bigger with time ✓
smaller with time ______
can't say ______

Explain your reason(s)

There's a divergent boundary in the middle of it pushing the plates away and creating new land

b.) Is the Pacific Ocean getting:
bigger with time ______
smaller with time ✓
can't say ______

Explain your reason(s)

The Pacific ocean is mainly a convergent boundary which means that the pacific plate will go under the Eurasan/Philippine plate, which gets rid of oceanic floor

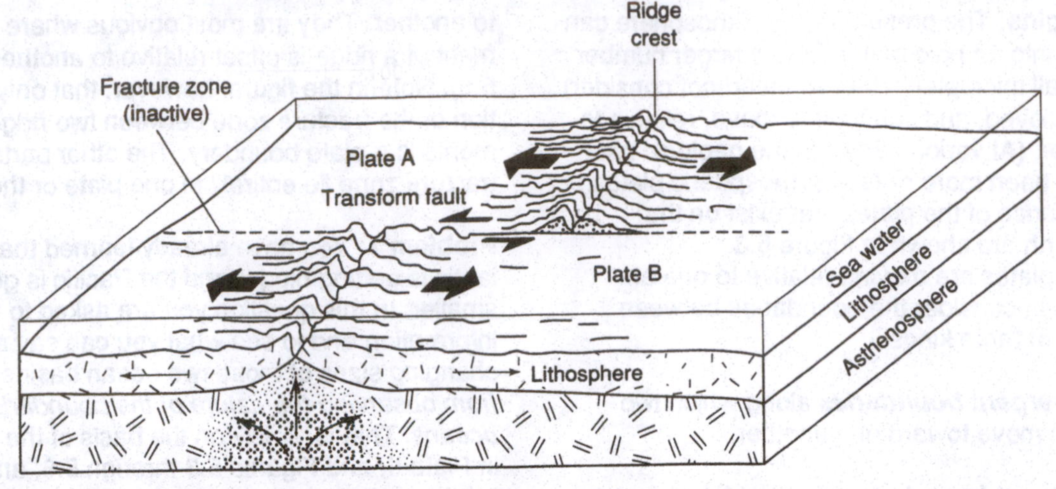

Figure 5.6. Transform boundaries are boundaries along which plates slide relative to one another. In this figure, the portion of the fracture zone between the two ridge segments is a transform boundary (or transform fault). The portions of the fracture zone that are not between the ridge segments lie entirely within a single plate and are therefore not plate boundaries. They are, however, scars of transform plate motion in the past. (Reprinted with the permission of Macmillan Publishing Company from *The Earth,* Fourth Edition, fig 18.22, by Edward J. Tarbuck and Frederick K. Lutgens. Copyright © 1993 by Macmillan Publishing Company.)

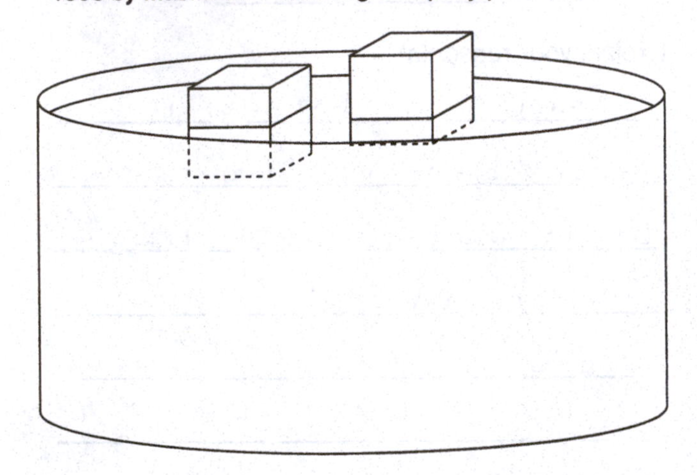

Figure 5.7. The denser a floating object is, the greater the fraction of it that will be submerged. In this figure, about 50 percent of the block on the left is emergent as is about 75 percent of the block on the right. The block on the left is, of course, more dense.

c.) Is the passage between Antarctica and Australia getting bigger or smaller?

bigger

Explain your reasons

There's a divergent boundary between the 2 pushing them away, and there's a convergent boundary on top of Australia pulling Australia closer to the Eurasian plate.

Problem 2: If abyssal plains are formed when sediment from the continents blankets topographic features of volcanic origin on the seafloor, why do you think that abyssal plains are much more widespread in the Atlantic (where trenches are not very common) than in the Pacific (where the continental margins are bordered by trenches in so many places)?

Since the Atlantic doesn't deal with convergent boundaries then it doesn't have as many trenches as the Pacific.

Trenches = Convergent

Isostasy and seafloor subsidence: The rocks of the earth's crust are of lower density than those of the underlying mantle. The lithosphere (the crust and the rigid upper part of the mantle) floats buoyantly upon the asthenosphere (the more plastic zone of rock that underlies it). This provides the basis for understanding why the spreading axes (the ridges and rises) are high and why the seafloor deepens away from the spreading axes.

Icebergs have a density about 90 percent of

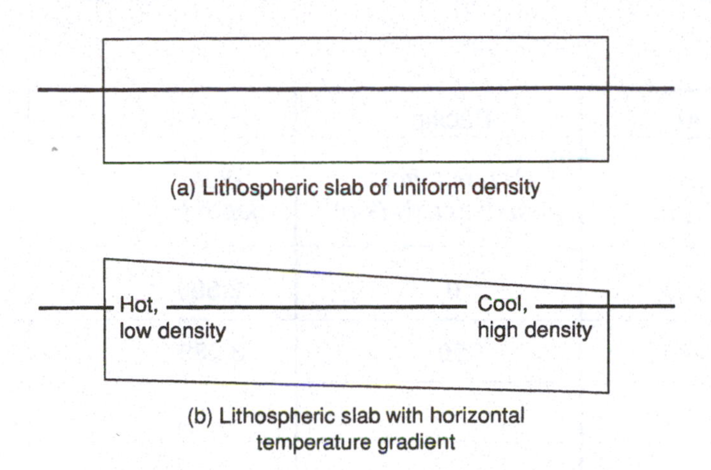

Figure 5.8. The surface of a floating slab of uniform thickness, temperature and density (upper sketch) is horizontal, because all parts of the slab are buoyed up to the same extent. If the left end of the same slab is heated while the right end is cooled (lower sketch), the surface of the slab will slant downward to the right. This is because the right end becomes denser. A greater fraction of that end will be submerged.

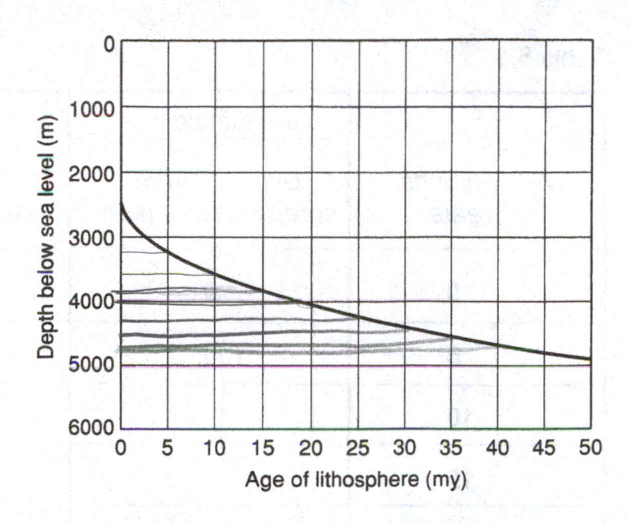

Figure 5.9. This curve shows the relationship between the age of normal oceanic lithosphere and the depth below sea level of the seafloor. This single curve describes the subsidence of the ocean floor in all of the oceans very well.

that of seawater. As a result, approximately 90 percent of a floating iceberg is below the surface of the ocean and ten percent is above. In general, the lower the density of a floating object, the higher it will float (Figure 5.7).

Imagine a slab of lithosphere floating buoyantly on the underlying asthenosphere. If the thickness, mass and density of the lithosphere are the same everywhere, the surface of the lithosphere will be horizontal, as shown in the upper sketch of Figure 5.8. But remember that rocks and almost everything else (except fresh water between 0° and 4°C) contract, or become denser, as they cool. Now consider a slab of lithosphere that is similar to the one we just discussed, except that its temperature decreases from a high value at the left to a low value on the right. The effect of this is shown in the lower half of Figure 5.8. The right end of the slab, because it is cooler, is more dense than the left end. A greater fraction of the right end is therefore submerged and the surface of the slab slants downward from left to right.

Now think of the seafloor spreading process at a divergent plate boundary (Figure 5.5). The lithosphere that is generated by the intrusion of hot, molten basaltic lava along the spreading axis has lower density than the older, cooler lithosphere farther from the plate axis. The result, as shown in Figure 5.8, is that the lithosphere floats higher close to the spreading axis. Older, cooler lithosphere sinks further down into the asthenosphere. This is why divergent plate boundaries are ridges and rises that stand higher than the surrounding abyssal plains.

Geophysicists are able to calculate the rate of cooling of a lithospheric plate as heat is carried away by the seawater bathing its upper surface. With this information it is possible to calculate the density of any part of the lithospheric plate and hence the level to which it subsides in the asthenosphere. These calculations are beyond the scope of an introductory course, but the results of the calculations are useful. They indicate that the temperature and the density of a section of normal oceanic lithosphere depend on the age of that section. In other words, if we know the age of a part of the ocean floor, we ought to be able to estimate the depth of the water above it. This is illustrated in Figure 5.9. The reverse is also true. If the water depth at a location is known, and if the oceanic lithosphere at that location is normal (i.e., not a seamount, a trench, etc.), it is possible to estimate the lithospheric age at that site.

Problem 3: A typical half-spreading rate along the Mid-Atlantic Ridge is 20 km per million years. Along the East Pacific Rise 150 km per million years is a typical value. Using this information and the data in Figure 5.9, fill in Table 5.1. Use the results for 5 million years ago as an example.

Now use the data in Table 5.1 to plot graphs for both the Mid-Atlantic Ridge and the East Pacific Rise showing water depth as a function of distance from the spreading axis. Plot the graphs on Figure 5.10. Label each of the curves. Remember that the topography on one side of the rise or ridge is the mirror image of the topography on the other side.

Table 5.1.

Age (millions of years)	Atlantic Distance from spreading axis (km)	Atlantic Water depth (m)	Pacific Distance from spreading axis (km)	Pacific Water depth (m)
0	0	2,500	0	2,500
5	100	3,280	750	3,289
10	200	3,500	1,500	3,500
15	300	3,800	2,250	3,800
20	400	4,000	3,000	4,000
25	500	4,300	3,750	4,300
30	600	4,400	4,500	4,400
35	700	4,600	5,250	4,600
40	800	4,700	6,000	4,700
45	900	4,800	6,750	4,800
50	1,000	4,900	7,500	4,900

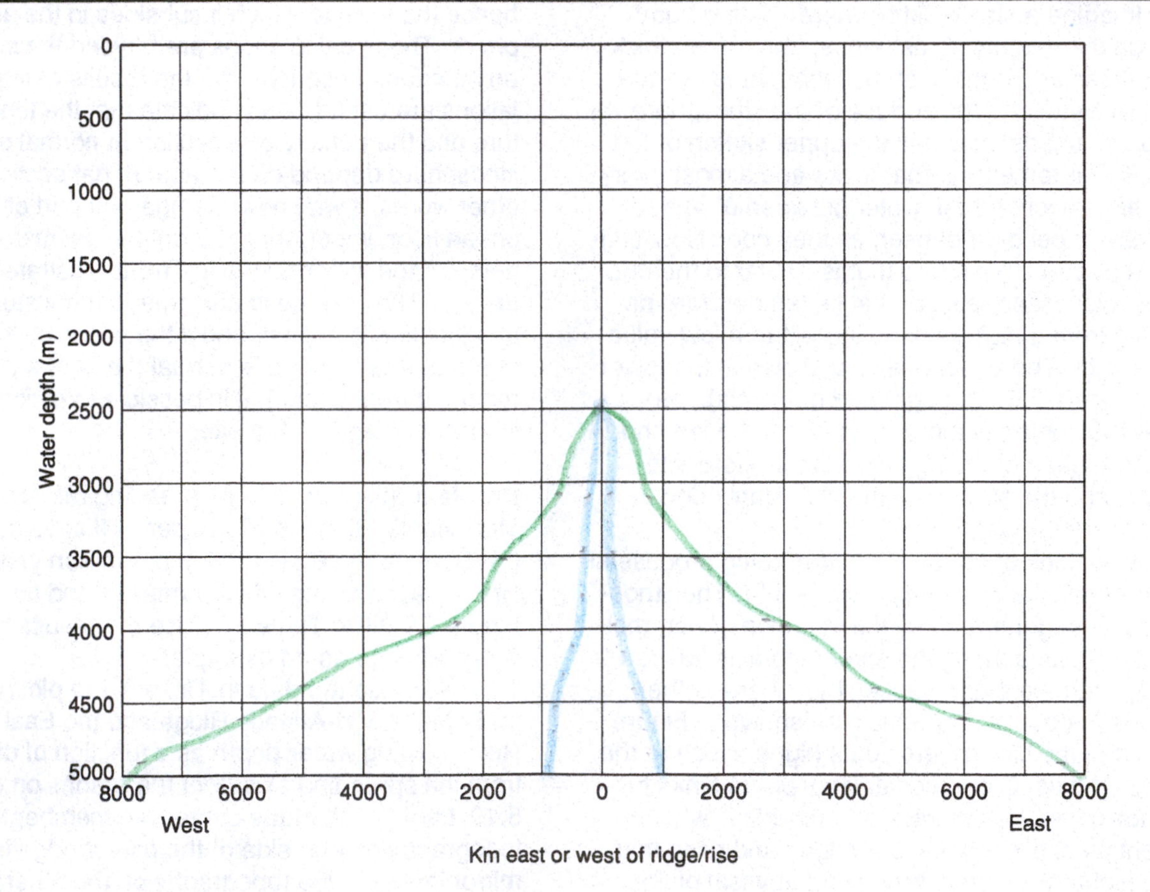

Figure 5.10. Plot your graph for problem 3 here.

Use your results in Figure 5.10 to describe the difference between the topography of the Mid-Atlantic Ridge and that of the East Pacific Rise.

The Pacific has a more gradual slope slowly gets deeper. The Atlantic gets deep very fast.

Now, make a general statement about the relationship between spreading rate and the topography of an oceanic ridge or rise.

The faster the spreading rate, the smaller the slope is. The slower the spreading rate the faster it gets deep.

Seismicity of plate boundaries: An earthquake ***epicenter*** is the location on the surface of the ground (or the floor of the ocean) directly overlying the spot within the earth at which the earthquake-generating motion occurred (the ***earthquake focus***). Epicenters of earthquakes that occurred over a period of several years are plotted in Figure 5.11. Compare the locations of these epicenters to the locations of plate boundaries seen in Figure 5.3. Note

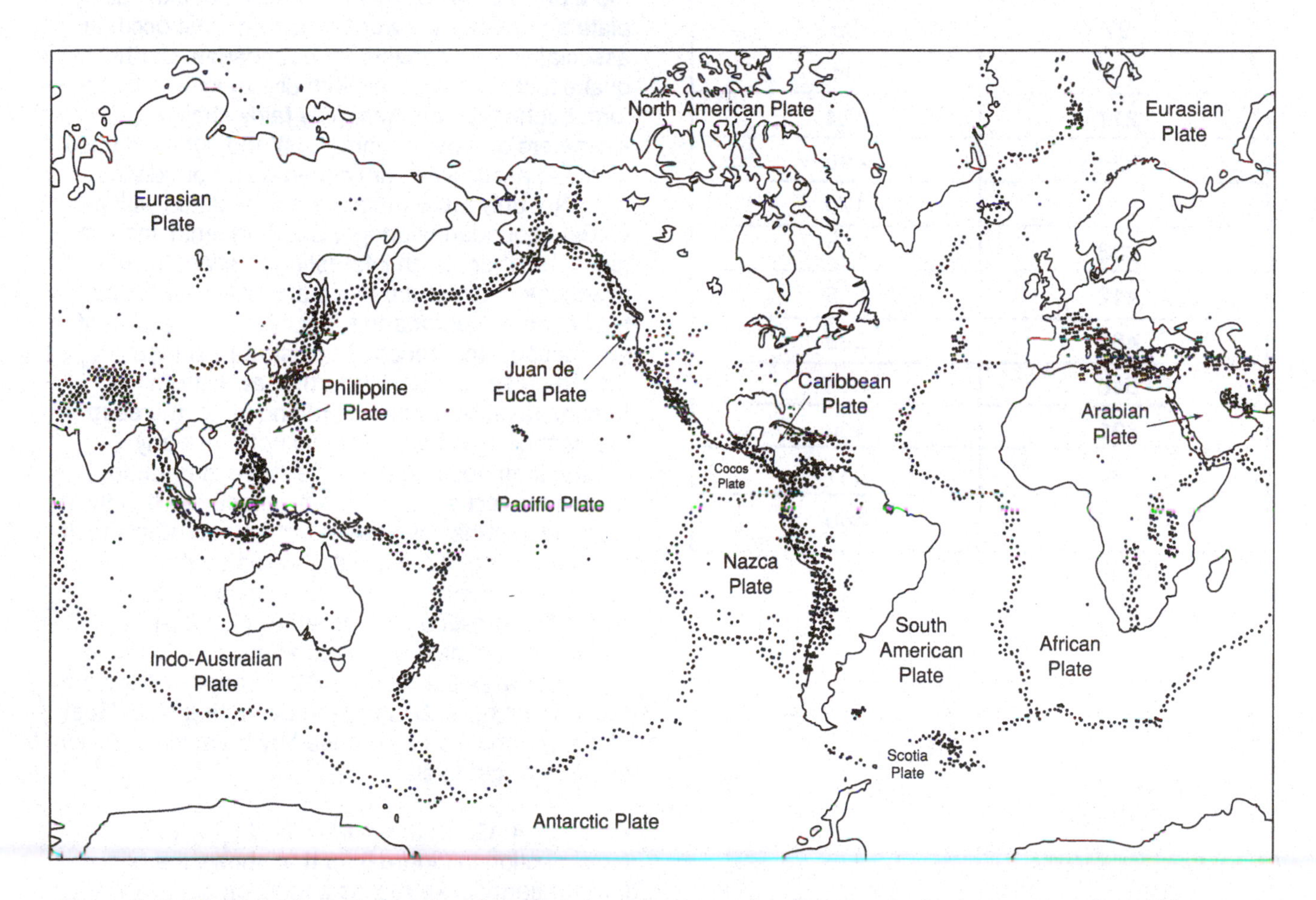

Figure 5.11. World distribution of earthquakes. Note that the pattern of earthquake distribution matches that of lithospheric plate boundaries. (Reprinted with permission of Macmillan Publishing Company from *Earth Science*, Sixth Edition, fig. 5.11, by Edward J. Tarbuck and Frederick K. Lutgens. Copyright © 1991 by Macmillan Publishing Company.)

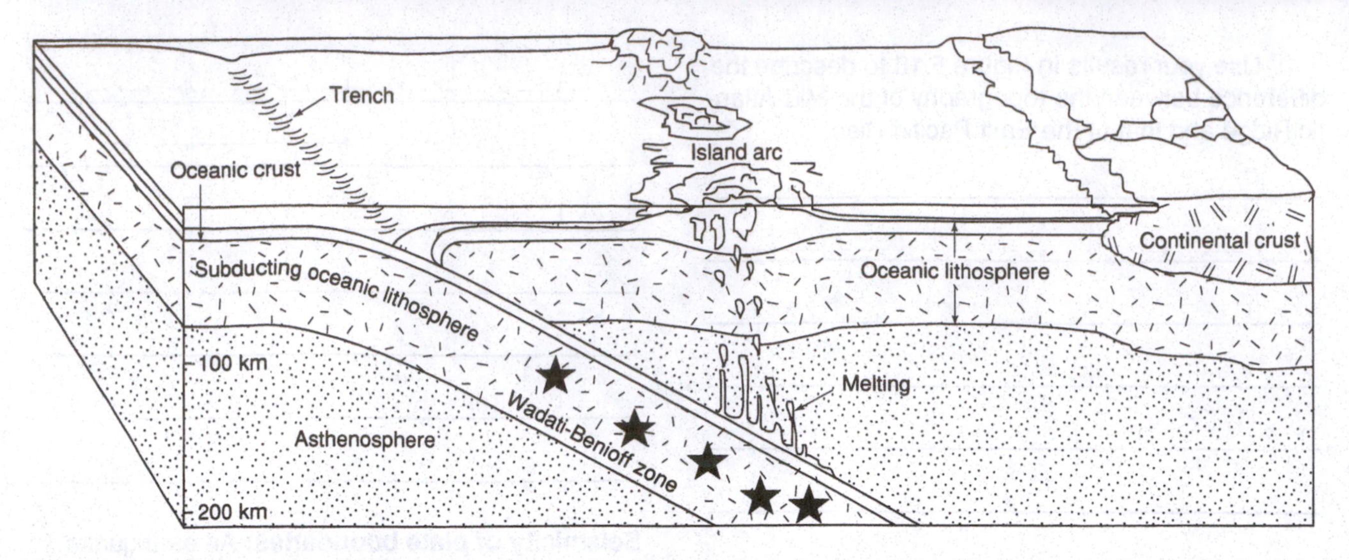

Figure 5.12. Relation between descending plate and earthquake foci, shown by stars. (Adapted with the permission of Macmillan Publishing Company from *Earth Science,* Sixth Edition, fig. 6.11, by Edward J. Tarbuck and Frederick K. Lutgens. Copyright © 1991 by Macmillan Publishing Company.)

Table 5.2.

Distance from Trench (km)	Depth to focus (km)
97	55
138	84
211	124
820	490
294	170
333	200
412	235
481	266
546	315
691	400
964	551
1008	600

that the majority of all earthquakes are associated with plate boundaries. (Some, it is true, do occur within the interiors of plates.) If we were to look in more detail at the distribution of earthquakes along plate boundaries we would find that most occur in association with convergent boundaries. Earthquake foci in association with divergent and transform boundaries are generally fairly shallow (a few kilometers or tens of kilometers) and not as intense as those associated with convergent boundaries.

Not only are earthquakes associated with convergent boundaries deeper and frequently more intense, but there is an interesting relationship between the depth of the focus and the distance between the epicenter and the surface expression of subduction—the trench. As illustrated in Figure 5.12, a cross section, the farther away from the trench, traveling on the overriding plate, the deeper the earthquake focus. (Foci do not occur any deeper than about 700 km, because at greater depths the rocks are sufficiently plastic so that they respond to stress by flowing rather than breaking. It is the breaking of rocks that causes earthquakes.) The zone in Figure 5.12 that contains the earthquake foci is called the Wadati-Benioff Zone, or sometimes simply the Benioff zone, after the seismologists who first described it. It marks the trace of the downgoing slab under the overriding slab. Most of the earthquakes are caused by breaking of rocks in the downgoing slab.

Problem 4: Given the data in Table 5.2, plot a cross section in Figure 5.13 that shows the positions of earthquake foci as a function of distance from the trench. Use a protractor to determine the

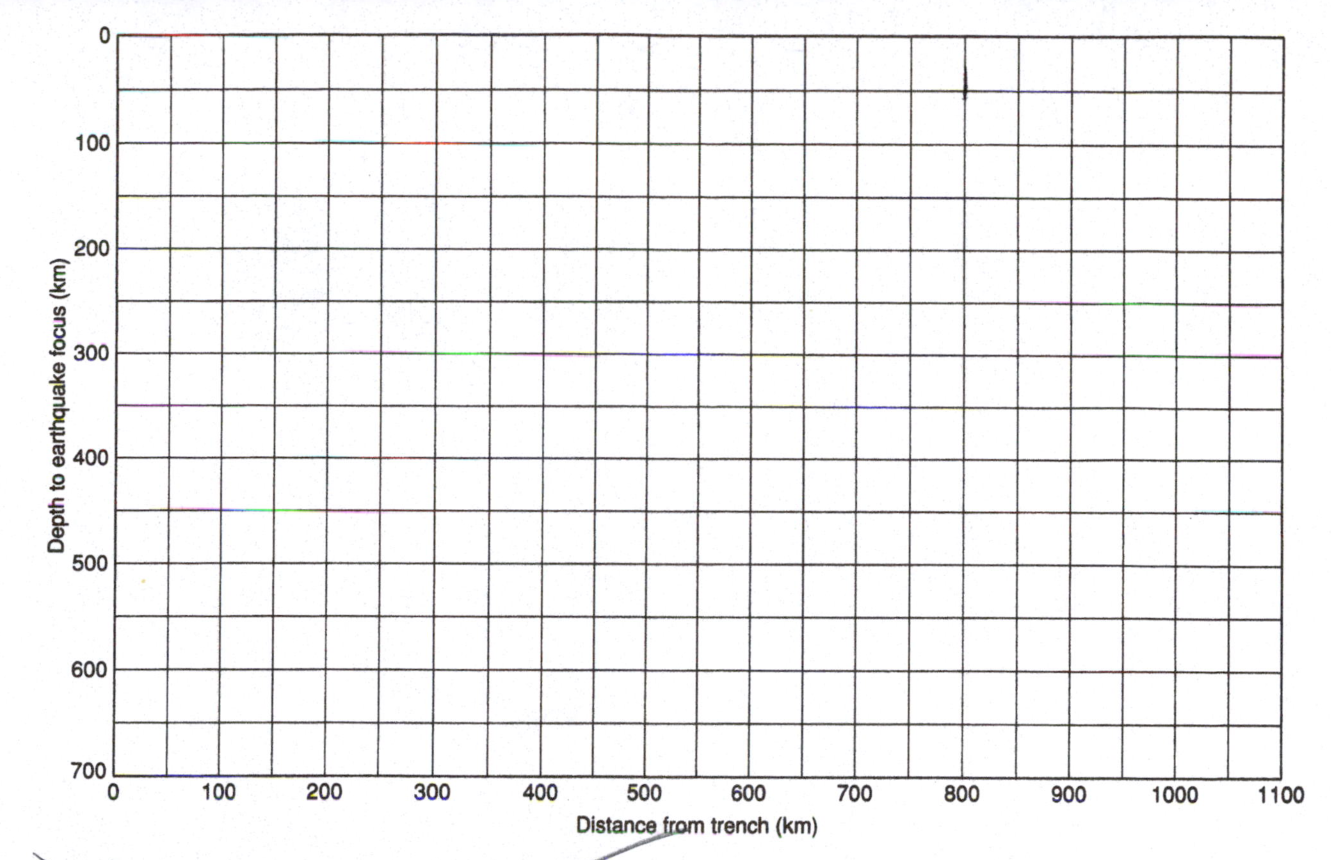

Figure 5.13. Plot the cross section of Problem 4 here.

approximate angle at which the downgoing slab is descending into the asthenosphere. (In determining the angle, consider a horizontal surface to have an angle, or dip, of 0° and a vertical surface to have a dip of 90°

At what angle is the downgoing slab descending?

____

Name: ______________________ Date: ____________ Section: ____________

Exercise 6
OCEAN SEDIMENTS

Purpose: Almost everywhere except in the vicinity of mid-ocean spreading centers, the floor of the ocean is covered with a layer of sediment. This sediment is composed of particles which may have originated either on the continents or within the oceans, and which may have been produced by biological processes, abiological processes, or a combination of the two. After the sedimentary particles are deposited on the floor of the ocean and become buried by particles deposited on top of them, they may undergo modifications. In some cases the sediments contain valuable natural resources. In this exercise we will examine some of the geological processes on the ocean floor, and the nature of the sediments that result.

Equipment and materials: The use of a set of sieves and some sand samples is optional. You will find a calculator to be useful in doing this exercise.

Marine sedimentary provinces: In our discussion of marine sediments, it is useful to distinguish between those sediments deposited on the continental margins and those deposited on the floor of the deep oceans. The continental margin is often referred to as the ***neritic environment*** and the sediments deposited upon it are ***neritic sediments***. Sediments deposited on the floor of the deep oceans may be referred to as ***oceanic sediments*** or ***pelagic sediments***. Not surprisingly, neritic sediments are usually much more strongly affected by processes and events on the continents than are pelagic sediments.

Origins of sedimentary components: It is useful, for many purposes, to categorize the origins of the components of marine sediments. (In this context, a component of a sediment can be thought of as all of the particles of the sediment that have a similar origin and composition.) ***Terrigenous*** (or ***lithogenous***) components originate from the chemical weathering and/or physical disintegration of rocks exposed on land. The particles may be carried to the oceans by rivers, by wind, or by glaciers. The terrigenous components consist of the large variety of rock-forming minerals found on the

A.

B.

Figure 6.1. The most common biogenous components of pelagic sediments are microscopic. They include radiolaria and foraminifera, shown in *A*, magnified approximately 160 times, and coccoliths, shown in *B*, magnified about 10,000 times. (Courtesy of Deep Sea Drilling Project, Scripps Institution of Oceanography, University of California, San Diego.)

Table 6.1. Definitions of particle size ranges according to the Wentworth Classification.

Size	*Particle*
>256 mm	Boulder
64 to 256 mm	Cobble
4 to 64 mm	Pebble
2 to 4 mm	Granule
2 to 1 mm	Very coarse sand
1 to 1/2 mm	Coarse sand
1/2 to 1/4 mm	Medium sand
1/2 to 1/8 mm	Fine sand
1/8 to 1/16 mm	Very fine sand
1/256 to 1/16 mm	Silt
<1/256 mm	Clay[a]

[a] The term *clay* is used in two different ways. In this table it is used to define a range of particle sizes. It is also used to describe a group of minerals, related to the micas and composed primarily of silicon, aluminum and oxygen. Most clay *minerals* are in the clay particle *size* range.

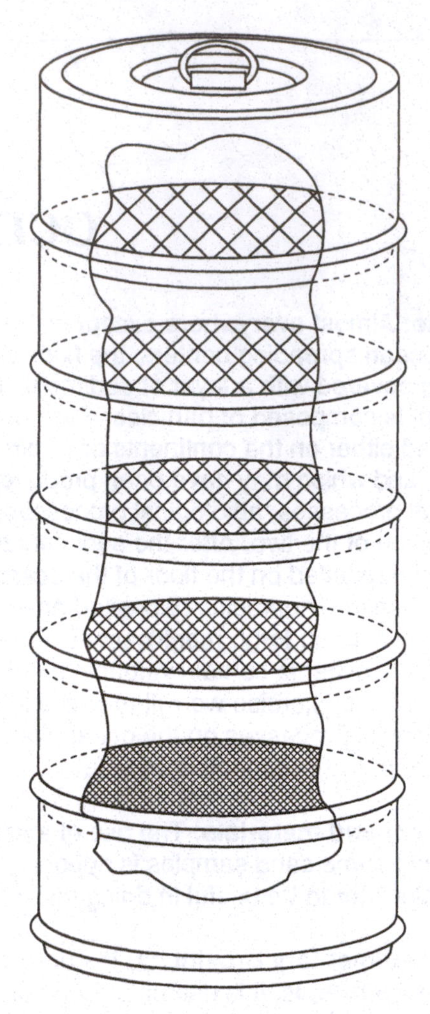

Figure 6.2. A stack of sieves used to determine the panicle size distribution of coarse-grained (sand and silt) sediment.

continents, including quartz (SiO_2), clay minerals and feldspars. ***Biogenous*** components are those that are produced by biological processes in the ocean. In the deep sea, most biogenous components are either calcium carbonate ($CaCO_3$) or opaline silica[1] ($SiO_2 \cdot nH_2O$) produced by single-celled plants and animals (Figure 6.1), but the remains of some higher organisms are also found. In coastal regions, a much larger variety of organisms produces biogenous sediments. Most spectacular of these sediments are the reefs produced by corals. A third major class of sedimentary components is ***hydrogenous*** components. These are formed by chemical reactions within seawater or between seawater and rocks or sediments. They include clay minerals formed by the alteration of submarine volcanic rocks, phosphate deposits, and manganese nodules. It is likely that bacterial processes are involved in the formation of some hydrogenous components, but the role of organisms is far less obvious than it is in the formation of biogenous components. ***Cosmogenous*** components, which originate in space, largely as micrometeorites, are present widely but in low abundance.

Particle size distributions of marine sediments: Sedimentary particles occur in a large range of sizes. The size of each particle reflects its origin and the manner in which it was transported to the site where it was deposited. Students who live in areas that have been glaciated (e.g., much of the northern United States and most of Canada) are familiar with large boulders, sometimes weighing many tons, that have been carried long distances by the ice sheets that covered those regions in the past. Ice is capable of carrying particles far larger than can moving water, and water can carry larger particles than can be carried by the wind.

It is often helpful, in trying to understand the

[1] Note that the chemical formula for opaline silica ($SiO_2 \cdot nH2O$) is very similar to that of the mineral quartz (SiO_2). Quartz is a ***crystalline*** substance, meaning that its constituent atoms are arranged in three dimensions in a regular, orderly fashion. Opaline silica, or hydrated, ***amorphous*** silica differs from quartz in that (1) its atoms are not arranged in an orderly fashion and (2) its structure contains variable amounts of water (signified in the formula as nH_2O, where n represents a number that can vary).

origins of the components of a sample of ocean sediment, to characterize the sediment in terms of its particle size distribution (Table 6.1). For coarse sediments (sand and silt) this may be readily done using a set of sieves (Figure 6.2). A dry sample of sediment is weighed and placed in the topmost (coarsest) sieve of the set. The set is then covered and shaken for a few minutes, and the amount of sediment in each of the sieves is then weighed. If no material has been lost, the total weight of the separate ***size fractions*** should equal the weight of the sediment at the start of the experiment.

As sediment is transported through the water, it tends to get sorted out according to particle size. Coarser particles generally settle out of suspension first. The factors that control the particle size distributions of sediments are complex, but sediment that has been carried far from its source[2] often has a unimodal distribution (see Problem 7, Exercise 2 for definition of unimodal and bimodal distributions). As the sediment becomes better sorted by transport over longer distances, each sample acquires a more narrowly defined size distribution. Sediments that have been transported further (and in more slowly moving water) are typically composed of finer particles than are sediments that are closer to their source (Figure 6.3).

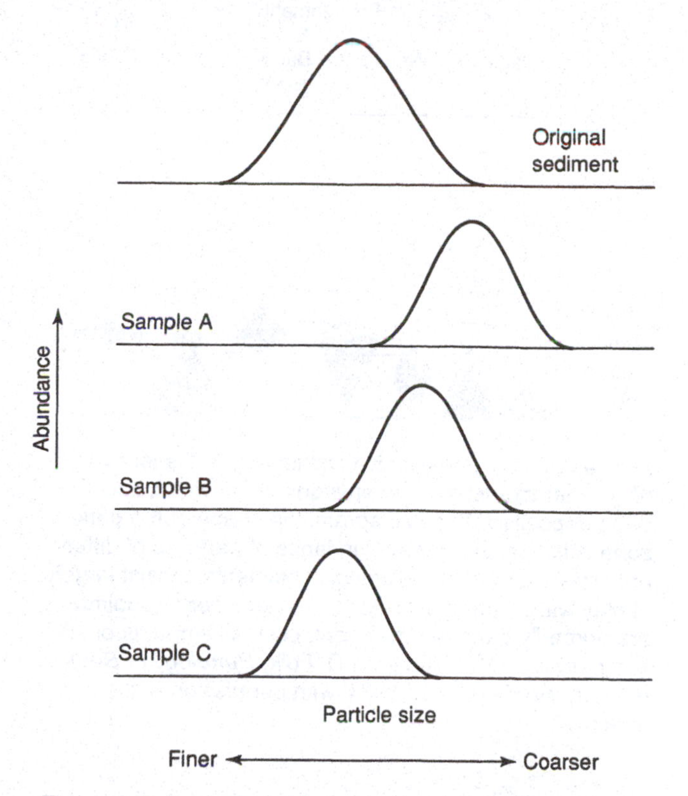

Figure 6.3. As sediment is transported from its source, it often becomes sorted according to particle size. This sketch shows the unimodal particle size distribution of a sediment (labeled *original sediment*) and the particle size distributions of three sediment samples from three different locations (labeled *A, B,* and *C*) all of which formed as the result of sorting of the original sediment during transportation.

If sediments from two or more sources, each with a unimodal particle size distribution, become mixed, the result will often be sediment with a ***bimodal*** or ***polymodal*** particle size distribution (Figure 6.4). A bimodal distribution may also result when two different processes are affecting the sediment. For example, river sediment may have a bimodal distribution if it consists of particles that have been transported in suspension in the water and other particles that have been transported by bouncing along the river bed.

The particle-size distribution of a sediment is often characteristic of the environment in which it has been deposited. Figure 6.5 illustrates, in a very general way, the typical coarsening of sediments as one goes from the deep sea to the continental shelves and then further landward to the beach. Figure 6.6 shows a similar coarsening of sediments found in many shore areas. You can often observe this coarsening at the beach as you walk away from the water toward the higher portions of the shore.

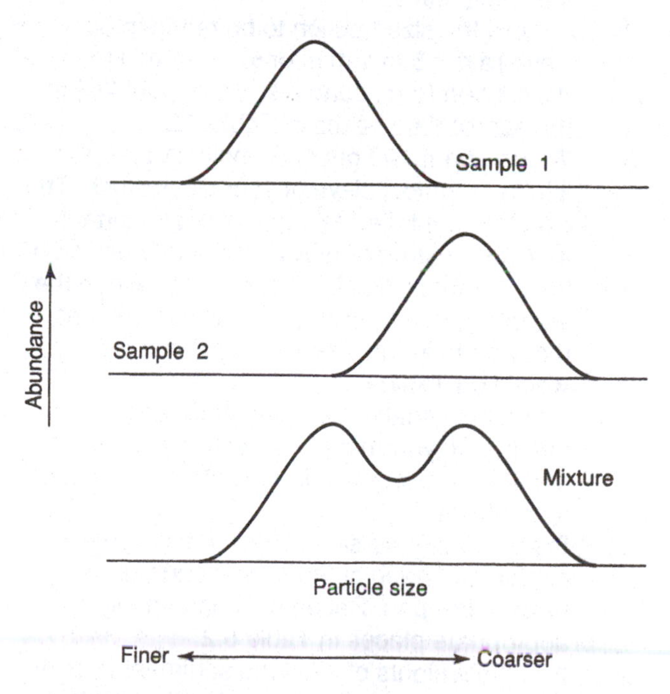

Figure 6.4. A sediment formed by the mixture of sediments from two different sources, each with a unimodal distribution, may have a bimodal distribution.

Problem 1: You may have been provided with sieves, two sediment samples, and a balance with

---

[2] It is not necessary for sediment to be transported in a straight line in order to become well-sorted. For example, sediment may become well-sorted as the result of being washed back and forth by waves.

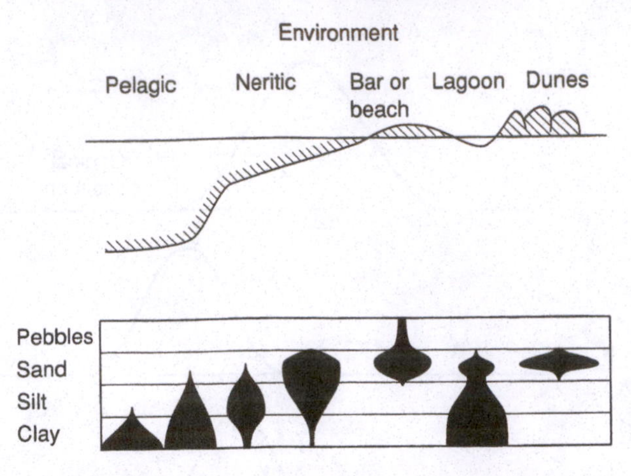

Figure 6.5. The particle size distribution of a sediment is often characteristic of the environment in which it has been deposited. In this diagram, the shape of the dark zone is indicative of the abundance of particles of different sizes. For example, pelagic sediments consist largely of clay with a small amount of silt, while beach sediments are primarily sand but may contain small amounts of silt and pebbles. (Modified from R. Folk, *Petrology of Sedimentary Rocks*, p. 107,1974, with permission of the author.)

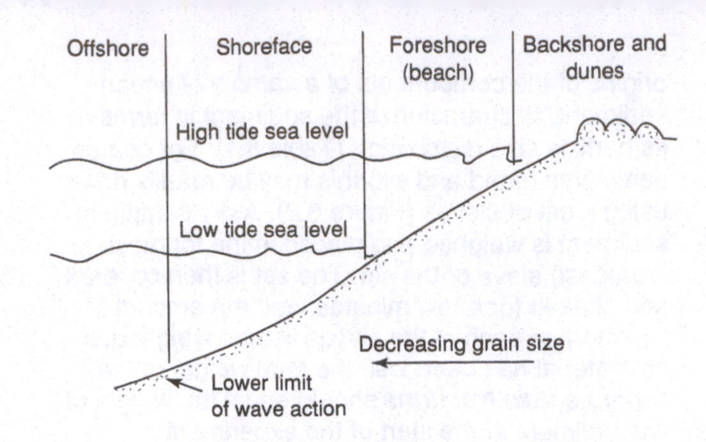

Figure 6.6. In the shore zone, sediments commonly become finer-grained in the offshore direction. (Adapted from M.E. Tucker, *Sedimentary Petrology—An Introduction*, Blackwell Scientific Publications, fig. 2.56, 1981.)

which to weigh the sediments and the fractions collected in the sieves. If so, proceed as follows. If not, go on to Problem 2.

1. Assemble your sieve stack with the coarsest sieve on the top and the finest on the bottom. The sieve stack should have a pan at the bottom to catch the sediment that passes through the finest sieve.
2. Record the size fraction to be retained by each sieve (e.g., .5 to 1 mm, or >2 mm) and the size description (e.g., coarse sand, or pebbles) in the appropriate boxes of Table 6.2.
3. Weigh about 100 gm of Sample #1 and place it in the topmost sieve of your sieve stack. The exact amount, i.e., 100 gm, is not important, and your instructor may direct you to use a different amount. But it is important to weigh the sample you are sieving carefully and record that weight in the space marked *starting weight* on Table 6.2.
4. Alternately shake the sieve stack and slap it gently with your hand. (Or, you may shake it and then tap it on a table.) Do this for three or four minutes.
5. Disassemble the sieve stack and weigh the contents of each of the sieves and the contents of the pan. Record this data in the appropriate places in Table 6.2.
6. Add the weights of all of the sediment in all of the sieves and in the pan and put the total in the blank in Table 6.2 marked *total of fractions*. Calculate the difference between the *starting weight* and the *total of fractions*. A large difference will indicate that you have lost some material or that you have made an error in weighing or tabulating data. Calculate what percentage of the total sample each fraction constitutes. The percent of the sample is given by:

$$\frac{\text{Weight of fraction}}{\text{Total of fractions}} \times 100$$

7. Clean your sieves. (Do not use water because it will take them too long to dry.) Repeat items 3 through 6 for Sample #2.

Problem 2: In this problem you are asked to draw histograms of particle size distribution. The instructions to follow depend upon whether you did Problem 1 and obtained your own particle size data. If you did Problem 1, use the data in Table 6.2 to draw a histogram of particle size distributions. To do this, copy from Table 6.2 to Figure 6.7a the size range and size description of each size fraction. Then draw a bar for each fraction of length corresponding to the percentage of that fraction. (For example, if a fraction constitutes 37 percent of the sample, shade the corresponding row on Figure 6.7 from 0 percent to 37 percent.) If you did not do Problem 1, use the data in Table 6.3. These data describe particle size distributions of two sediment samples collected from the coastal zone of Florida (Visher, 1969). First calculate the percentage of the sample that each size fraction constitutes, and enter those values in the appropriate boxes of Table 6.3. Then plot your histogram as described in the preceding paragraph, using Figure 6.7b.

Table 6.2. Data sheet for Problem 1.

	SAMPLE	#1		#2	
	NAME				
Size Range	Size Description	Weight (gm)	Percent of Sample	Weight (gm)	Percent of Sample

Total of fractions	gm	pct	gm	pct
Starting weight	gm		gm	
Difference between total of fractions and starting weight	gm		gm	

Problem 3: Describe the particle size distribution you plotted for each sample in Figure 6.7. Indicate, for each sample, whether the particle size distribution is unimodal or bimodal.

Sample 1:
This sample is unimodal b/c it only has one peak at fine sand.

Sample 2:
This sample is bimodal b/c it peaks at 2 points which are coarse sand & fine sand.

Table 6.3. Data describing the particle size distributions of two sediments from the coastal zone of Florida. Use this table for Problems 2 and 3 if you were not instructed to do Problem 1.

	SAMPLE		#1	#2	
Size Range	Size Description	Weight (gm)	Percent of Sample	Weight (gm)	Percent of Sample
>1 mm	Very coarse sand	0.00		0.00	
0.84 to 1 mm	Coarse sand	0.58	.5%	7.30	10
0.71 to 0.84 mm	Coarse sand	0.35	.3%	1.10	1.5
0.59 to 0.71 mm	Coarse sand	0.11	.1%	2.56	3.5
0.50 to 0.59 mm	Coarse sand	0.34	.3%	3.65	5
0.42 to 0.50 mm	Medium sand	0.69	.6%	2.19	3
0.35 to 0.42 mm	Medium sand	1.15	1%	3.65	5
0.30 to 0.35 mm	Medium sand	2.53	2.2%	2.92	4
0.25 to 0.30 mm	Medium sand	2.30	2%	5.84	8
0.21 to 0.25 mm	Fine sand	4.60	4%	5.84	8
0.18 to 0.21 mm	Fine sand	11.50	10%	9.49	13
0.15 to 0.18 mm	Fine sand	21.85	19%	8.00	10.96
0.13 to 0.15 mm	Fine sand	36.80	32%	8.50	11.6
0.11 to 0.13 mm	Very fine sand	24.15	21%	8.03	11
0.09 to 0.11 mm	Very fine sand	4.60	4%	3.29	4.5
0.07 to 0.09 mm	Very fine sand	2.88	2.5%	0.95	1.3
0.06 to 0.07 mm	Very fine sand	0.52	.5%	0.07	0.096
<0.06 mm	Silt	0.05	.04%	0.07	0.096

Total of fractions		115.00 gm	pct	73.00 gm	pct

Source: Glenn S. Visher, Grain size distributions and depositional processes, *Journal of Sedimentary Petrology*, vol. 39, pp. 1074–1106 (1969).

Which sample has a narrower particle size distribution? Which sample do you think has been more effectively sorted with respect to particle size as the result of sedimentary processes? Explain your reasoning.

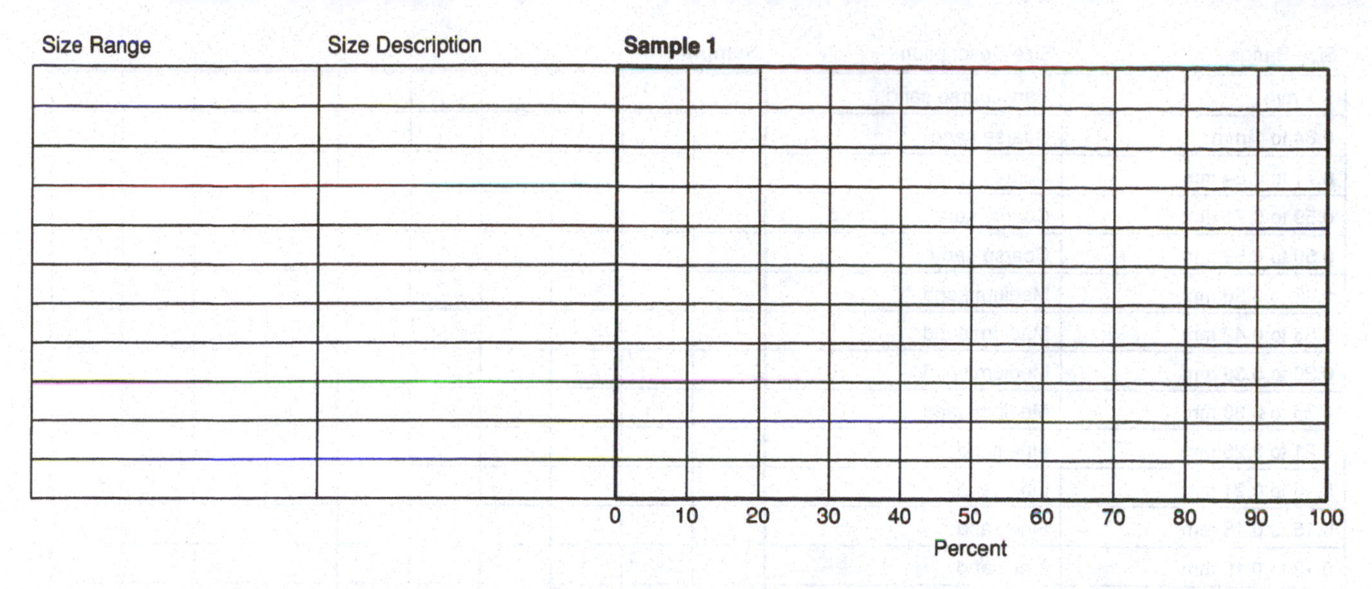

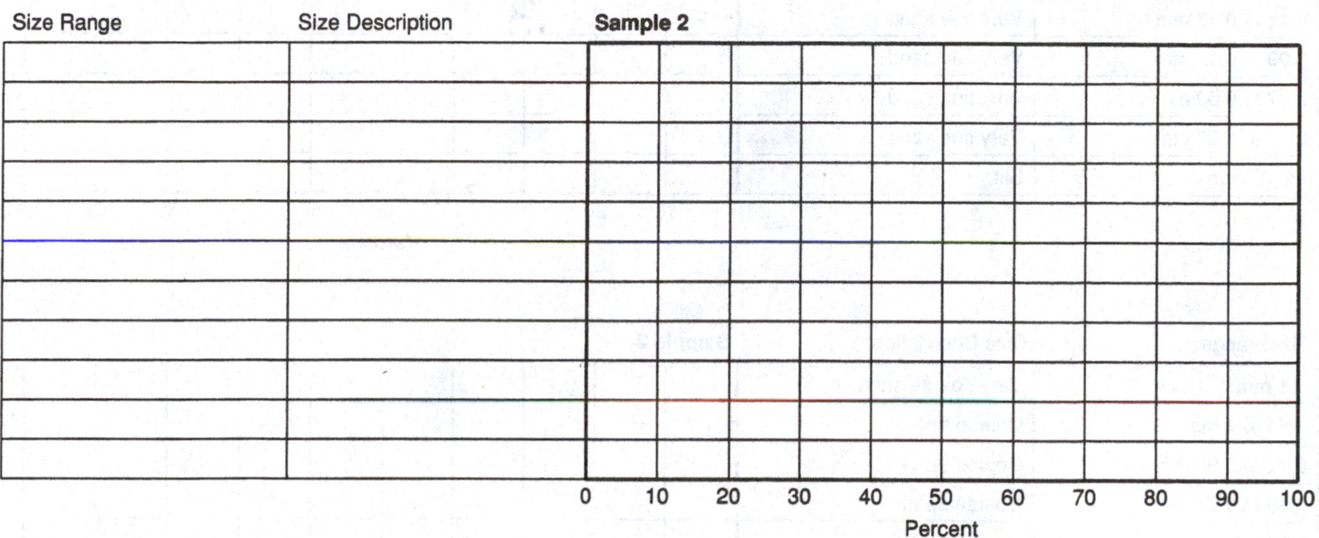

Figure 6.7a. Use this figure to draw the histograms required in Problem 2 if you are using the particle size data you obtained and tabulated in Table 6.2.

Sample 1 has a narrower particle size distribution because it's just mainly fine sand
Sample 2 was more sorted through the sedimentary process.

Turbidites and turbidity currents: The continental rise at the base of the continental slope (Exercise 5) often consists largely of sediment that has been transported down the slope from the continental shelf. Among the more spectacular mechanisms by which sediment can be transported down the slope to the rise is a ***turbidity current***. A turbidity current is a dense slurry of sediment suspended in water, capable of moving rapidly downslope along the bottom of the ocean or lake. It may be generated when an earthquake or other disturbance causes sediment on a steep slope to become suspended in the overlying water. The suspension, which flows as if it were a fluid much more dense than seawater, can accelerate rapidly as it moves down the slope. It is capable of eroding sediment over which it travels, thereby increasing its size as it flows. It may flow until it reaches the continental rise and abyssal plain where the gentle slopes cause it to slow down, allowing the sediment to settle out. The sediment deposited at this point, called a ***turbidite***, is characterized by ***graded bedding***. That is, the sediment becomes progressively more fine-grained in the upward direction. This is because the coarser particles settle out before the finer particles.

Sample 1

Size Range	Size Description
> 1 mm	Very coarse sand
0.84 to 1 mm	Coarse sand
0.71 to 0.84 mm	Coarse sand
0.59 to 0.71 mm	Coarse sand
0.50 to 0.59 mm	Coarse sand
0.42 to 0.50 mm	Medium sand
0.35 to 0.42 mm	Medium sand
0.30 to 0.35 mm	Medium sand
0.25 to 0.30 mm	Medium sand
0.21 to 0.25 mm	Fine sand
0.18 to 0.21 mm	Fine sand
0.15 to 0.18 mm	Fine sand
0.13 to 0.15 mm	Fine sand
0.11 to 0.13 mm	Very fine sand
0.09 to 0.11 mm	Very fine sand
0.07 to 0.09 mm	Very fine sand
0.06 to 0.07 mm	Very fine sand
< 0.06 mm	Silt

0 10 20 30 40 50 60 70 80 90 100

Percent

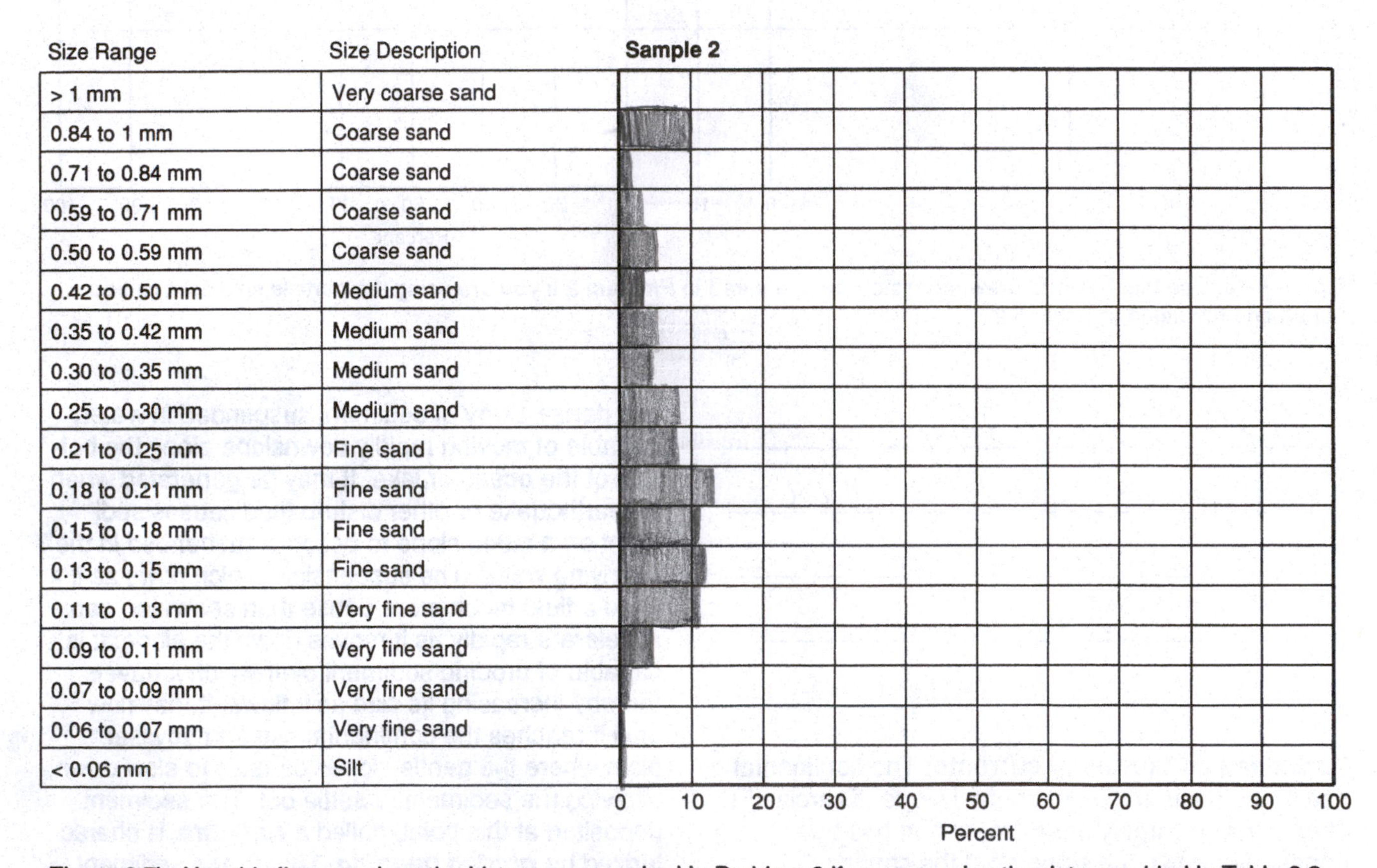

Sample 2

Size Range	Size Description
> 1 mm	Very coarse sand
0.84 to 1 mm	Coarse sand
0.71 to 0.84 mm	Coarse sand
0.59 to 0.71 mm	Coarse sand
0.50 to 0.59 mm	Coarse sand
0.42 to 0.50 mm	Medium sand
0.35 to 0.42 mm	Medium sand
0.30 to 0.35 mm	Medium sand
0.25 to 0.30 mm	Medium sand
0.21 to 0.25 mm	Fine sand
0.18 to 0.21 mm	Fine sand
0.15 to 0.18 mm	Fine sand
0.13 to 0.15 mm	Fine sand
0.11 to 0.13 mm	Very fine sand
0.09 to 0.11 mm	Very fine sand
0.07 to 0.09 mm	Very fine sand
0.06 to 0.07 mm	Very fine sand
< 0.06 mm	Silt

Figure 6.7b. Use this figure to draw the histograms required in Problem 2 if you are using the data provided in Table 6.3 for samples from the coastal zone of Florida.

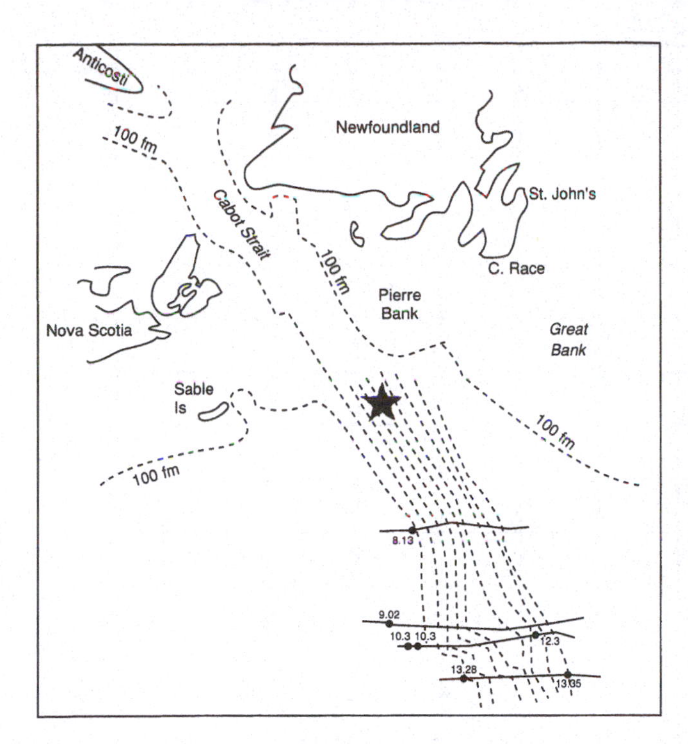

Figure 6.8. Map of the Grand Banks, Nova Scotia region, showing the locations of cable breaks following the 1929 earthquake. The star shows the location of the epicenter. The straight lines show the locations of cables that broke. The vertical ticks on the straight lines are locations of breaks. (Reprinted with permission from *Nature,* Volume 124, J.W. Gregory, The earthquake south of Newfoundland and submarine canyons, Pages 945–946, Copyright 1929, Macmillan Magazines Limited.)

Table 6.4. Data concerning the telegraph cable breaks caused by the Grand Banks earthquake of November 1929.

Break #	Distance from epicenter (km)	Water depth (m)	Hours after quake
1	277	4387	8.13
2	450	5484	9.02
3	508	5831	10.30
4	510	5301	10.30
5	551	5118	12.30
6	591	5082	13.28
7	655	5118	13.35

Probably the best documented turbidity current to have occurred in the oceans is one that took place in November 1929 on the Grand Banks, south of Newfoundland (Figure 6.8). The region was crossed by a number of Transatlantic telegraph cables. Within an hour after the occurrence of an earthquake centered on the Grand Banks, the cables began to fail in a regular sequence. The cables that were shallowest and closest to the earthquake epicenter failed first. Cables further down the slope failed progressively, as depth and distance from the epicenter increased. The location of each break could be determined by electrical measurements. The time at which each break occurred was accurately known by the telegraph company that operated the cable, because it was the time when telegraph transmission on that cable was interrupted. Each break occurred where the cable crossed a submarine canyon. When the broken cables were retrieved for repair they were found to be covered with turbidite sediments. The relationship between the time of occurrence and the location of the epicenter, the locations and times of the cable breaks, and the geologic environment provide convincing evidence that the breaks were the result of a turbidity current triggered by the earthquake.

Problem 4: The distance from the earthquake epicenter to each cable break is listed in Table 6.4. The time elapsed between the earthquake and each cable break is also tabulated. Use that data to plot, on Figure 6.9, the distance from the epicenter of the Grand Banks earthquake to each cable and the interval between the earthquake and the cable break. Draw a straight line that passes through the origin (i.e., the point for 0 km at 0 hr) and best fits the data you have plotted. Use that line to calculate the average rate of travel of the turbidity current responsible for the breaks. (Remember that velocity = distance travelled/time elapsed.)

What was the average speed of the Grand Banks turbidity current of 1929 in km/hr? Using the conversion table in Exercise 1, convert that value to statute miles per hour.

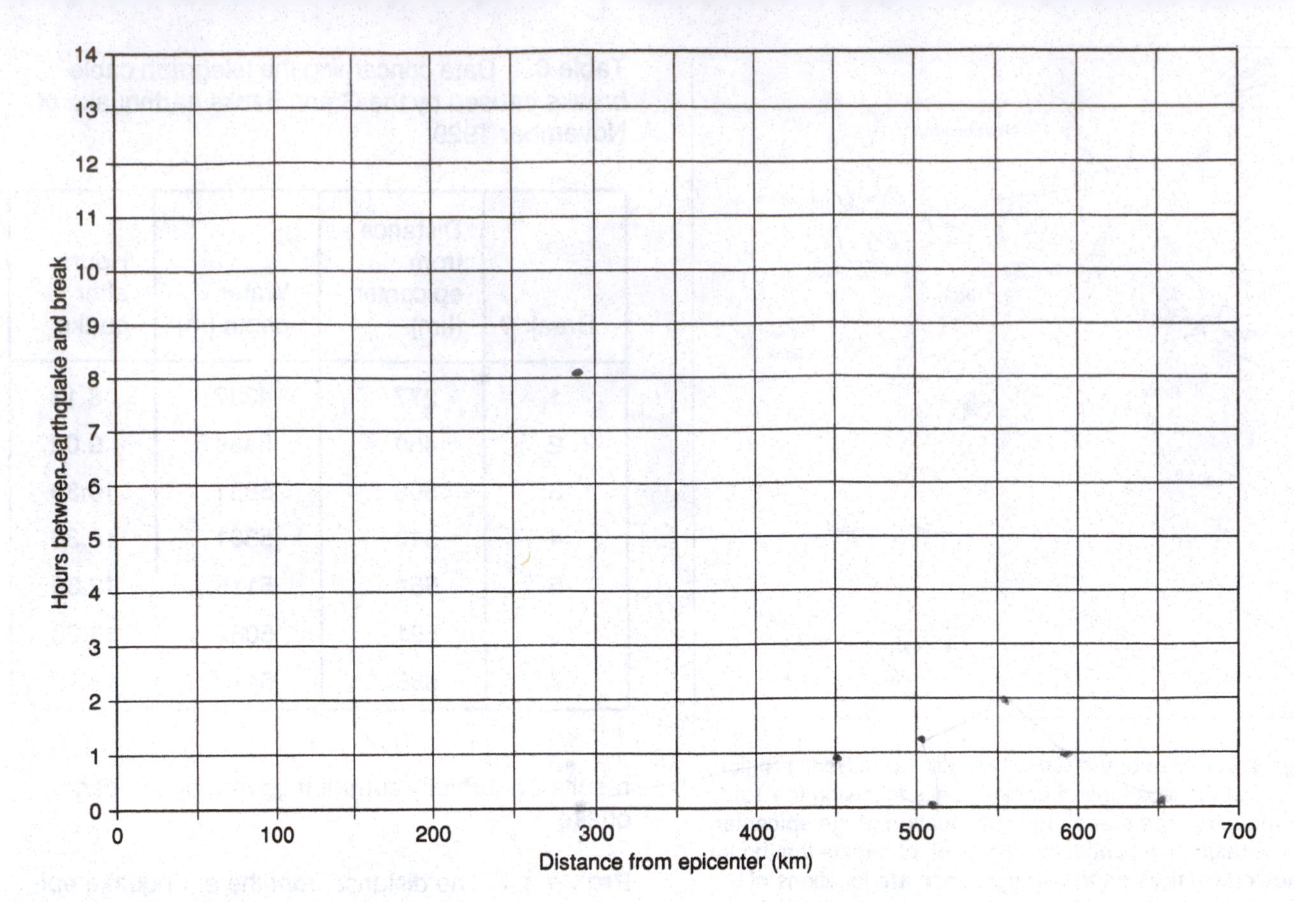

Figure 6.9. Plot the data for Problem 4 here.

Name: ______________________ Date: ______________ Section: ____________

Exercise 7
CHEMICAL ASPECTS OF SEAWATER

Purpose: Even before you began to study oceanography you probably knew a number of facts about the chemistry of seawater. You undoubtedly knew that seawater is salty and that the main constituent of the salt is sodium chloride, NaCl, or common table salt. But if you have swum in the ocean and tasted the water, you probably noticed that it didn't taste quite like a solution of table salt in water. It was more bitter. Seawater is, in fact, a complex solution of many chemicals. The origins of different ***solutes***, or dissolved substances, in seawater are different, and the processes that govern the concentration and distribution of each one presents an interesting story. In this exercise you will examine the origins and the processes that determine the concentrations of some of the constituents of seawater.

Equipment and materials: A calculator will be helpful but is not necessary. For Problem 4 you will need a Styrofoam coffee cup or other container in which you can make a small hole, a 250 cm³ (or larger) graduated cylinder, a watch or timer capable of measuring seconds, a ruler at least 15 cm (6 inches) long that is graduated in centimeters, and a sink or other source of running water.

The origin of the H_2O of the hydrosphere: In the simplest view, seawater is a mixture of two components, H_2O (which makes up about 96.5 percent of seawater), and the material dissolved in it (which makes up the rest). As you do this exercise, interpret the terms *water* or *seawater* to mean the solution of the H_2O plus the solutes.

There is good reason to believe that most of the H_2O of the hydrosphere escaped to the surface of the earth from the interior of the planet fairly early in the earth's history. This process of transport of water from the earth's mantle to the surface is called ***outgassing***. We can see a great deal of water coming from the interior to the surface of the earth today, especially in the form of hot springs, and geysers and volcanic emanations. Almost all of this water, however, originated as rainwater that percolated downward into the crust and reacted with hot rocks before returning to the surface of the earth. But a small amount of the water coming to the surface of the earth today (and undoubtedly much more when the earth was a young planet) is ***juvenile water***, or water that is reaching the surface of the earth for the first time.

Geologists are generally in agreement that the earth formed about 4.5 billion years ago as the result of the agglomeration of meteorites. Most meteorites consist either of an alloy of iron and nickel, similar in composition to the core of the earth, or of aluminosilicate minerals, similar to those that make up the earth's mantle and parts of the crust. The latter are called ***stony meteorites***. There are many different types of stony meteorites but many contain small amounts of water or contain minerals that liberate water when they are heated.

Problem 1: In this problem you are to estimate the amount of H_2O that might have existed in the mantle of the earth early in its history[1]. You are then to compare that amount with the amount of H_2O in the oceans. Finally you will conclude whether or not the H_2O of the oceans could have originated by outgassing of the earth's interior.

First calculate the volume of the earth's mantle. A cross section of the earth is shown in Figure 7.1. (In the figure, and in this problem, the crust is ignored. The volume of the crust is tiny in comparison with the volume of either the mantle or the core.) Remember that the volume of a sphere is given by the formula

$$V = \frac{4}{3} \pi r^3$$

where π has a value of 3.14 and r is the radius of the sphere. Use the data in Figure 7.1 to calculate the volume of the mantle. *Remember that the earth and the core are spheres but the mantle is not!* The

[1] In this problem we consider the mantle early in the history of the earth, but not during the very earliest history. It is likely that in the earliest history of the earth the planetary interior was a mixture of particles derived from stony meteorites and particles derived from iron meteorites. The earth probably became *differentiated*, or separated into a core and a mantle, some time in the first billion years of its history.

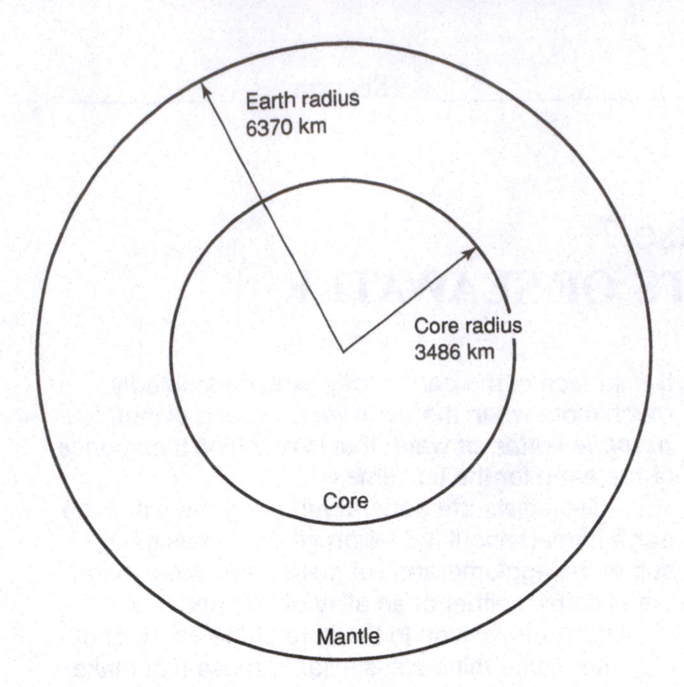

Figure 7.1. A cross section through the earth. The volume of the crust is very small in comparison with either the core or the mantle, and is ignored in this figure.

mantle is that part of the earth that is not core, so volume of the mantle is

$$V_{mantle} = V_{earth} - V_{core}$$

If you do the calculation using radii expressed in km, the answer you get will be in km^3 (cubic kilometers). You need to express the answer in cm^3 (cubic centimeters). Remember that 1 $km^3 = 10^{15}$ cm^3. If you have forgotten how to express large numbers in scientific notation, review Exercise 1.

Show your calculation here. Express the answer in cm^3.

Now, calculate the mass of the mantle. Express your answer in grams. Assume that density of the mantle is 4.5 gm/cm^3 (about 4.5 times the density of H_2O). Remember that

$$\text{Mass (g)} = \text{Volume (cm}^3\text{)} \times \text{Density} \left(\frac{\text{gm}}{\text{cm}^3}\right)$$

Next, assume that the mantle is made up of material from stony meteorites. Stony meteorites contain about 0.5 percent (by weight) H_2O. Calculate the mass of water in the mantle when it was formed from the meteorites. Express your answer in grams.

Now calculate the volume of the oceans. To do this, take the surface area of the oceans to be 3.6×10^8 km^2. Take the average depth to be 3.8 km. Calculate the volume using the formula

$$\text{Volume (cm}^3\text{)} = \text{Area (cm}^2\text{)} \times \text{Height (cm)}$$

or

$$\text{Volume (km}^3\text{)} = \text{Area (km}^2\text{)} \times \text{Height (km)}$$

Remember that

$$1 \text{ km} = 10^5 \text{ cm}$$

so

$$1 \text{ km}^3 = (10^5 \text{ cm})^3 = 10^{15} \text{ cm}^3$$

Express the answer in cm^3.

Next calculate the mass of H_2O in the oceans. To do this, assume that the density of seawater is 1.025 gm/cm³ and that seawater is 96.5 percent H_2O. Express the answer in grams.

Finally compare the mass of H_2O in the oceans to the mass of H_2O originally contained in the mantle. Which is bigger? By how much? Could the H_2O of the oceans have come from the outgassing of the mantle?

The salt of seawater: When we talk about the salt in seawater we are using the term in a way that is somewhat inaccurate. To the chemist, a salt is a solid[2] chemical compound composed of positive and negative ***ions*** (or electrically charged atoms). NaCl is a salt composed of positive sodium ions (Na^+) and negative chloride ions (Cl^-). The ions in the salt crystal are arranged adjacent to one another. But when the salt dissolves in water it dissociates. The ions become dispersed through the solution and move about randomly and quite independently from one another. If we make a solution of two salts, say NaCl and magnesium sulfate ($MgSO_4$) the resulting solution will contain four kinds of ions (Na^+, Cl^-, Mg^{++} and $SO_4^=$). It is not possible to distinguish by examining the solution that the Na^+ and Cl^- came from one salt and the Mg^{++} and $SO_4^=$ came from another. In other words, we cannot distinguish between a solution made of equal amounts[3] of NaCl and $MgSO_4$ from one that was made of equal amounts of Na_2SO_4 and $MgCl_2$. Thus, when we consider the chemical composition of seawater it is more appropriate to consider the concentrations of the various dissolved ions than the concentrations of various salts that one might imagine dissolving to make seawater.

The most abundant ions in seawater, and their concentrations, are listed in Table 7.1. Note that more than 99 percent of the dissolved salt in seawater is made up of only six ions. Also note that the percentages of the salt made up by each of the ions could be specified by only a single number, not a range of numbers. If we take seawater samples from all over the world and determine the abundances of the dissolved ions, we will always find that very close to 30.61 percent of the salt is Na^+ and very close to 1.16 percent is Ca^{++}. The total concentration of salt in the water (the ***salinity***) varies from sample to sample, but the relative proportions of the major ions that make up the salt do not.

During the early systematic exploration of the oceans in the late 19th and early 20th centuries it was difficult to measure the total concentration of salt in a seawater sample in a shipboard laboratory. But a straightforward technique was available, and readily usable on shipboard, for measuring the concentration of Cl^-. Oceanographers defined the chlorinity of a sample as the grams of Cl^- in 1 kg (1000 g) of seawater. The units of chlorinity are permil (‰), or parts Cl^- per 1000 parts seawater. Because the relative proportions of the major ions in seawater are constant, one can calculate the salinity (the total amount of salt in 1 kg of seawater, also expressed in units of permil) from the chlorinity.

Problem 2: A seawater sample has a chlorinity of 19.2‰. Use the data in Table 7.1 to determine its salinity.

[2] At high temperatures, in the absence of water, salts can, in fact, melt, but this need not concern us here.

[3] Equal amounts in this sense refers to ions with equal numbers of positive and negative electrical charges.

Table 7.1. Major Constituents of Seawater

Ion	*Percent*[*]
Chloride (Cl^-)	55.04
Sodium (Na^+)	30.61
Sulfate ($SO_4^=$)	7.68
Magnesium (Mg^{++})	3.69
Calcium (Ca^{++})	1.16
Potassium (K^+)	1.10
	99.28

[*]There are a number of ways to calculate the percentage of an ion in a solution. The numbers here are weight percents, or grams of each ion per 100 grams of salt.

Examining the effect of a tanker spill: During the summer of 1993 a tanker with a cargo of 10,000 tonnes (1 tonne is 1,000 kg or 10^6 grams) of sulfuric acid (H_2SO_4) foundered off the Pacific coast of Mexico and began to leak the acid into the ocean. When sulfuric acid dissolves in water it dissociates to form positive ions (H^+) and negative ions ($SO_4^=$). H^+ is the ion that gives acids their acidic properties. Its fate in a mishap like this one is somewhat complex. We will consider the effect of the spill only on the $SO_4^=$ concentration of the surrounding seawater. Normal seawater (salinity = $35^o/_{oo}$) contains 2.69 g $SO_4^=$ per kg of seawater.

$$\frac{35 \text{ g salt}}{1 \text{ kg water}} \times \frac{7.68 \text{ g } SO_4^=}{100 \text{ g salt}} = \frac{2.69 \text{ g } SO_4^=}{1 \text{ kg water}}$$

Problem 3: Imagine that the entire tanker cargo of H_2SO_4 were to be spilled suddenly into the surrounding ocean. In this problem you will do some calculations to help you estimate the impact of that spill on the environment. These calculations are outlined in detail. They are to be done in the spaces provided, *and the results are to be tabulated in the appropriate boxes of Table 7.2.*

a.) First, calculate the total amount of $SO_4^=$ (in grams) on board the tanker. (To do this, you need to determine how many grams of $SO_4^=$ are in 10,000 tonnes of sulfuric acid. It is important to know that sulfuric acid consists 98 percent of $SO_4^=$. In other words, each gram of sulfuric acid contains 0.98 grams of $SO_4^=$). Express your results in scientific notation.

The remainder of the problem deals with two different cases. In Case #1 the H_2SO_4 mixes into the uppermost 100 m of the water column and, as time passes, it spreads out laterally. Calculate the concentration in seawater of $SO_4^=$ resulting from the spill when it has mixed into a volume of water 100 m (approximately 300 ft) on a side and 100 m thick. This calculation requires several steps, each of which is described below.

b.) First, calculate the volume (in cm^3) of the seawater into which the H_2SO_4 mixes. This volume is a rectangular solid 100 m on a side and 100 m thick.

c.) Now calculate the mass (in kg) of the seawater in the volume you calculated above. Assume that the density of the water is 1.025 g/cm^3. Remember that[4]

$$\text{Mass (kg)} = \text{Volume (cm}^3) \times \text{Density } \left(\frac{\text{gm}}{\text{cm}^3}\right) \times \frac{1 \text{ kg}}{10^3 \text{ g}}$$

[4] Note that this formula differs slightly from the one you were given in Problem 1. In problem 1 you were asked to calculate the mass in gm. In this problem you calculate mass in kg. This accounts for the factor of 1/10^3.

d.) Next, calculate the concentration of SO_4^- in the contaminated volume of seawater that is derived from the spill (in g SO_4^-/kg seawater). Compare that concentration to the natural concentration of SO_4^- in seawater.

e.) Finally, determine the fraction of the SO_4^- in the contaminated volume of seawater that is derived from the spill and the fraction that is naturally occurring.

Now repeat the above calculations (b. through e.) for Case #2, in which the H_2SO_4 has mixed into a volume of seawater 1 km on a side and 100 m thick. Again, do the calculations in the space provided and put the results in the appropriate column of Table 7.2.

Table 7.2. Data sheet for Problem 3.

		Case #1	Case #2
1	Natural concentration of seawater SO_4^- in g SO_4^-/kg seawater (from discussion on page 62)	$\frac{2.69 \text{ g } SO_4^-}{\text{kg seawater}}$	$\frac{2.69 \text{ g } SO_4^-}{\text{kg seawater}}$
2	Weight of spilled SO_4^- in g (from 3a)		
3	Dimensions of volume of sea water	length 100 m width 100 m depth 100 m	length 1 km width 1 km depth 100 m
4	Seawater volume in cm^3 (from 3b)		
5	Seawater weight in kg (from 3c)		
6	Concentration of SO_4^- derived from spill in g SO_4^-/kg seawater (from 3d)		
7	Fraction of SO_4^- in contaminated volume that is derived from the spill (from rows 1 and 6)		

Volume of seawater (in cm^3):

____

____

____

____

Mass of seawater (in kg):

____

____

____

____

Concentration of SO_4^- in contaminated seawater that is derived from spill (in g SO_4^-/kg seawater):

____

____

____

____

Fraction of SO_4^- in contaminated seawater that is derived from spill:

____

____

____

____

It is sometimes said, although perhaps not completely fairly or wisely, that "the solution to pollution is dilution."

Steady state concentrations of salts: The dissolved constituents of seawater are constantly arriving in the ocean. River water commonly contains most of the common ions of seawater (although in different proportions), and some ions also become dissolved in the ocean when seawater reacts with hot rocks in the region of mid-ocean spreading centers[5] (Exercise 5).

In spite of the constant supply to the oceans, the concentrations of these solutes are not increasing with time. The reason for this is that other processes cause the solutes to be removed from seawater. Most important among these processes are the incorporation of dissolved materials in the sediments that accumulate on the bottom of the ocean. Sometimes this is quite obvious. Large volumes of calcium carbonate ($CaCO_3$) and silica ($SiO_2 \cdot nH_2O$) sediments are produced by organisms that remove dissolved Ca^{++}, CO_3^- and H_4SiO_4 from seawater. Some seawater is trapped in sediments as they accumulate. A portion of this leaks back into the overlying ocean as the sediments become compacted, but some may be removed from the oceanic system when sediments are subducted.

If the rate at which a solute is added to the ocean is exactly equal to the rate at which it is removed, the concentration of that solute does not change with time. The system is said to be in ***steady state*** with respect to that solute. The ocean is in steady state with respect to all major solutes and most minor ones. It might seem to be an almost magical coincidence that the rate at which each solute enters the ocean is identical to the rate at which that solute leaves the ocean, but it is in fact a rather straightforward consequence of a simple idea that you can readily test for yourself.

Take a Styrofoam coffee cup (or other container of similar size) and with a sharp pencil or ball point pen punch a hole in the bottom about 1/8 inch (3 mm) in diameter, as shown in Figure 7.2.

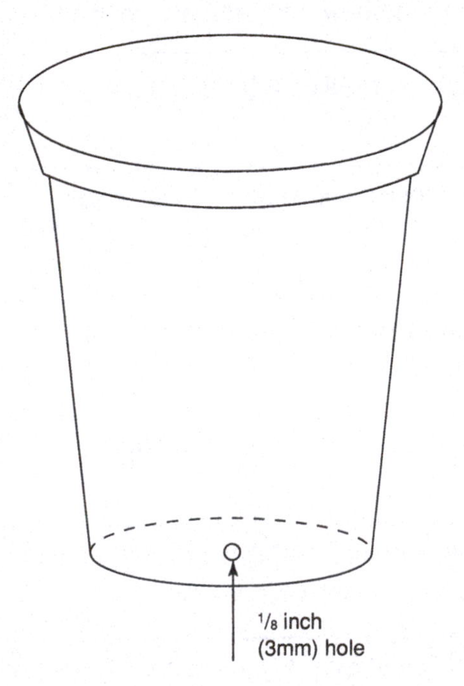

Figure 7.2. Punch a small hole in the bottom of a Styrofoam coffee cup.

[5] Some ions, most notably Mg^{++}, are removed from seawater by this process.

At a sink, turn on the water to a low but steady flow rate. Hold the cup under the stream of water and watch what happens to the water level inside the cup. The water will start to rise, and as it does, the rate of water flowing out of the hole in the cup will increase. Eventually the water level will stabilize. Water will flow out of the hole at the same rate it is flowing into the cup. The amount of water in the cup will be in steady state. (If the level rises to the point where it overflows the top you must slow down the rate of flow from the faucet. If the flow is as slow as it can go and the cup continues to overflow, you must make the hole in the bottom of the cup a little bigger.)

The amount of water in the cup reaches steady state because the rate of inflow from the faucet is constant. However, the rate of outflow through the hole in the bottom is not constant. The higher the water level in the cup, the greater will be the rate of outflow through the hole. That is because when the level of water is higher, the water pressure at the hole is greater. Because the pressure is greater, the water flows out faster. It can be shown mathematically (although we will not do so here) that whenever the rate of inflow to a system is constant and the rate of outflow varies in proportion to the amount of material in the system, the system will reach steady state.

Problem 4: Adjust the rate of inflow to the coffee cup or other container so that the water level reaches steady state when the cup is about 1/4 full. Verify that you have reached steady state by watching to see that the water level does not change for at least 15 seconds. Let the outflow go into the graduated cylinder and record, in the appropriate box in Table 7.3, the number of seconds necessary for 250 cm^3 of water to flow out of the cup. Next, hold your finger over the hole to stop the outflow and simultaneously shut off the water. Measure the height of the water surface above the bottom of the cup and record that value. Then, empty the graduated cylinder and use it to measure the amount of water in the cup. Record the volume. Repeat this experiment twice more, adjusting the rate of water inflow so that the steady state level is approximately 1/2 full, and again so it is approximately 3/4 full. (The more water in the cup, the longer you must watch it to verify that it has reached steady state. When the cup is 3/4 full you should be sure that the water level remains constant for at least a minute. Plot a graph in Figure 7.3 of the outflow rate vs. the height of the water above the bottom of the cup.

Note that the details of your results for Problem 4 will probably differ from those of other students doing the same experiment. That is because your results depend in part upon the size and shape of the hole you made in the bottom of the cup.

In words, describe the relationship you plotted in Figure 7.3 between the rate of outflow and the amount of water in the cup at steady state.

____

____

____

____

Returning to the concentrations of ions in seawater, the same principles that you saw to be operative in the cup of water hold for dissolved species in the ocean. Over long periods of time the rate of input to the oceans of most dissolved ions remains relatively constant. The rate of output of

Table 7.3. Record the measurements you make in Problem 4 here.

Fraction of cup filled	*Seconds for 250 cm^3 outflow*	*Outflow rate in cm^3/sec*	*Height of water above bottom of cup (cm)*	*Volume of water in cup (cm3)*

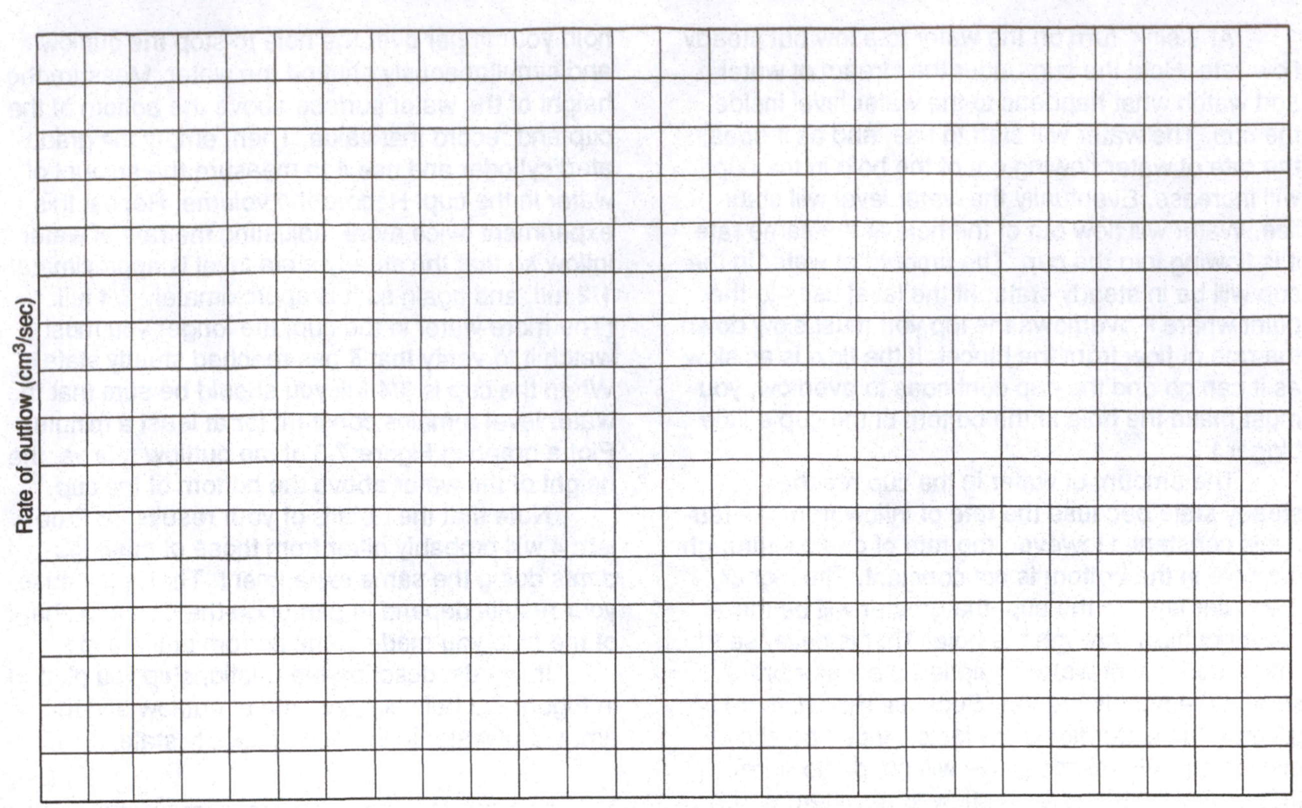

Figure 7.3. Plot a graph of the results of Problem 4 here.

many ions, however, is proportional to the amount of the ion dissolved in the ocean. For example, the rate of accumulation of calcium carbonate sediment on the floor of the ocean (and hence the rate of removal of dissolved Ca^{++} from seawater) may be expected to become greater as the concentration of Ca^{++} in seawater increases. As the concentration of any ion increases, the rate of its removal in water trapped in subducting sediments will increase. Thus the concentrations of most dissolved ions in seawater will tend to approach steady state.

Residence time: How long does an ion entering the ocean remain dissolved in seawater before it is removed by being incorporated into the sediments or the hot volcanic rocks in the vicinity of spreading centers? We can gain some insight into this by considering again the experiment of filling the coffee cup with the hole in the bottom. Imagine a fantasy world in which each drop of water entering the cup at the top left through the hole at the bottom in the same order that it entered. In this fantasy, a drop of water could not leave the cup until all of the drops that had entered before it had left. The length of time required for that to happen is given by

$$\frac{\text{Volume of water in cup}}{\text{Rate at which water leaves cup}}$$

For example, if the cup contained 150 cm^3 of water and the rate of outflow were 15 cm^3/sec, it would take 10 seconds for all of the water in the cup to be replaced.

In a real cup, of course, drops of water are not constrained to leave through the hole in the bottom in the same sequence they entered the cup. As a drop enters the cup it mixes with all of the other water, and different parts of the drop flow out at different times. But it still makes sense to think about the average time that water remains in the cup. The average time a drop (or a molecule) remains in the cup is called the residence time. Numerically, the residence time is identical to the length of time a drop would stay in the cup if drops were constrained to leave in the order they entered. That is,

$$\text{Residence time} = \frac{\text{Volume in cup}}{\text{Outflow rate}}$$

Since the system reached steady state, the outflow rate is identical to the inflow rate, so we can also write

$$\text{Residence time} = \frac{\text{Volume in cup}}{\text{Inflow rate}}$$

Problem 5: Calculate the residence time of water in the cup in each of the three experiments you did in

Problem 4. Put the calculations and the answers in the appropriate boxes of Table 7.4.

Residence times in the ocean: The residence time of a dissolved ion in seawater is the average length of time elapsed from when the ion enters the ocean to when it leaves seawater and becomes incorporated in the sediments or the volcanic crust. This is a useful concept in oceanography for a number of reasons. As you can see from the results of Problems 4 and 5, a system approaches quite close to steady state in three or four residence times. So the concentration of a seawater solute that has a residence time of 100 years is determined by processes that have occurred in the oceans over the past few hundred years. If oceanic conditions, input rates, and output rates have not changed very much over that time (and they probably have not) the ocean would be in steady state with respect to that solute. The concentration of a solute with a residence time of 100 million (10^8) years is determined by processes that have occurred over several hundred million years. There is a greater probability that a solute with a long residence time is not in steady state, but the rate at which we can expect its concentration to change would be extremely slow compared to a human lifetime. We might not be able to detect a departure of that solute from steady state. Using similar reasoning, we can see that natural processes will cleanse a polluted system more quickly if the pollutant has a short residence time than if it has a long residence time.

Problem 6: The amounts of several solutes in the oceans, and estimates of the rates at which they are entering or leaving the ocean are listed in Table 7.5. For each element listed, calculate the residence time and enter it in the appropriate box of the table.

Table 7.4. Put your answers to Problem 5 here.

Fraction of cup filled	*Residence time (seconds)*

Table 7.5. Computation sheet for the calculation of residence times of elements in seawater, based upon data in Broecker and Peng (1982).

Element	*Amount in ocean (g)*	*Rate of input or output (g/yr)*	*Residence time (yr)*
Sodium	151×10^{20}	1.8×10^{14}	
Potassium	5.6×10^{20}	4.6×10^{13}	
Calcium	5.8×10^{20}	5.2×10^{14}	
Silicon	3.9×10^{18}	3.0×10^{11}	
Manganese	3.9×10^{14}	3.0×10^{11}	
Iron	1.0×10^{14}	1.9×10^{12}	
Aluminum	1.1×10^{15}	1.8×10^{12}	

Name: ______________________ Date: ____________ Section: __________

Exercise 8

TEMPERATURE-SALINITY-DENSITY RELATIONS AND DENSITY STRATIFICATION

Purpose: In this exercise you will gain familiarity with the relations among the temperature, salinity and density of seawater and the reasons why seawater moves (or sometimes does not move) in the vertical direction. This understanding is essential to the understanding of the circulation of the oceans and, to some extent, of their biological productivity.

Temperature-density relationships of fresh water: Most substances contract (i.e., become more dense) as they cool. Fresh liquid water shares that behavior over most of its temperature range. However, in the temperature range 0° to 4°C, water expands (i.e., becomes less dense) as it cools. The density of liquid water as a function of temperature is given in Table 8.1.

Problem 1: On the graph paper provided in Figure 8.1, plot the density of water as a function of temperature. That is, plot the data in the table and draw a smooth curve through the points.

Table 8.1. Density of fresh water

Temperature (°C)	*Density (gm/cm³)*
0	0.999868
1	0.999927
2	0.999968
3	0.999992
4	1.000000
5	0.999992
10	0.999728
15	0.999127
16	0.998970
17	0.998802
18	0.998623
19	0.998433
20	0.998232
23	0.997567
25	0.997074

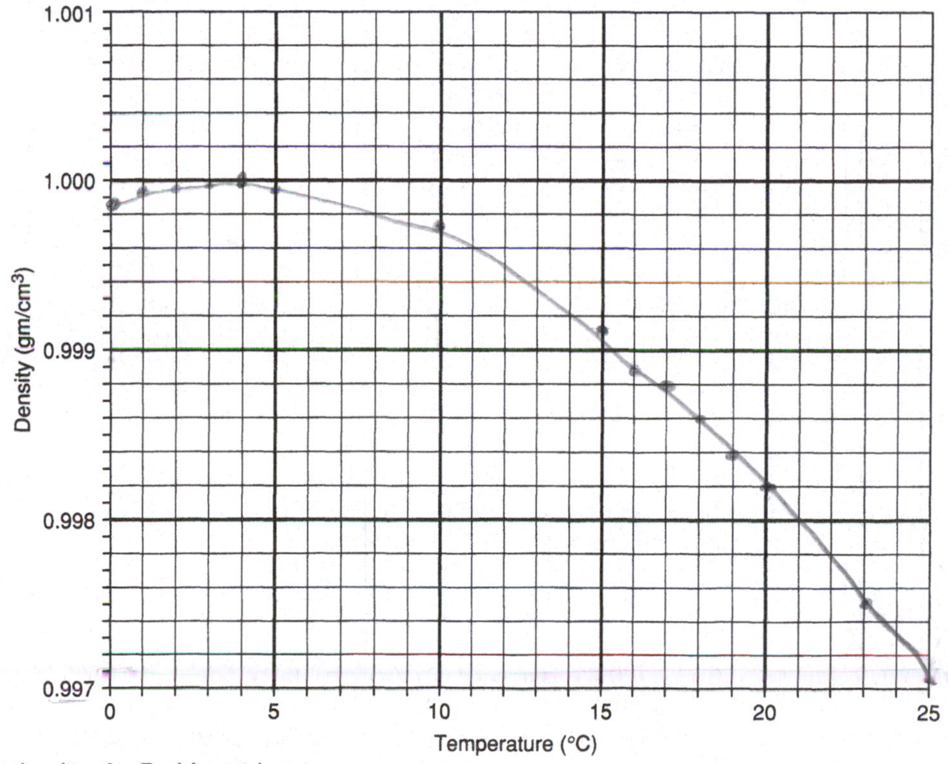

Figure 8.1. Plot the data for Problem 1 here.

Bathythermograms: A bathythermogram is a graph of water temperature vs. water depth. Ordinarily it is drawn with temperature along the horizontal axis and depth along the vertical, with the surface (depth = 0) at the top of the graph.

Water density and water depth: One of the most fundamental observations of fluids that we can make is that *less dense things always float on more dense things.* Oil floats on water. Cream floats on milk. Imagine that we introduce into the bottom of a lake or ocean some water that is less dense than the surrounding water. The water column will be unstable. The water we introduce will be buoyant. It will immediately begin to float upward until it lies above any water that is more dense.

Problem 2: In Figure 8.2 are given three bathythermograms such as you might measure in a deep, temperate lake in the winter, summer, and spring or fall. Using that data in Table 8.1 and/or the graph you drew in Problem 1, plot a density vs. depth curve corresponding to each of the bathythermograms.

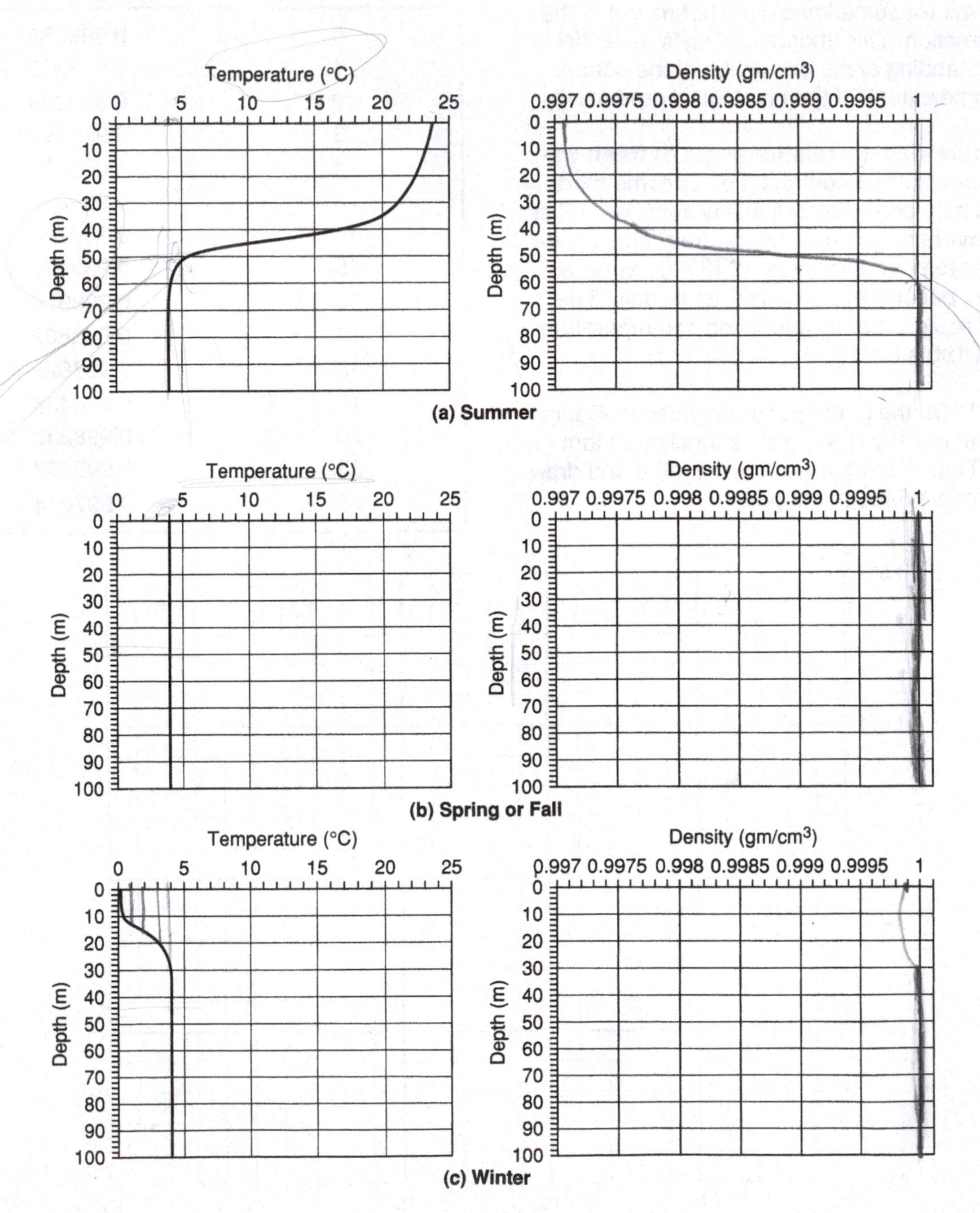

Figure 8.2. The graphs on the left are bathythermograms of a deep lake in a cool temperate climate during: (a) the summer; (b) the spring or fall; (c) the winter. The empty graphs on the right are to be used in the solution of Problem 2.

Problem 3: If the density of fresh water did not have a maximum at 4°C, but instead increased with decreasing temperature over the entire liquid water temperature range, it would be impossible to obtain a temperature-depth profile like that shown for the winter. Explain why this is so.

It wouldn't be possible to determine the temperature-depth profile because density has to cap at 1 and if it didn't the depth would go off the chart.

Salinity of seawater: The relative proportions of the dissolved major constituents of seawater (i.e., the salt) are almost the same everywhere in the oceans. This observation is sometimes referred to as the *law of constant proportions,* although it is not really a law, simply an observation. Another way of viewing the composition of seawater is to note that if we were to take seawater samples from all over the world and dry them out to remove the water there would be very little difference in the composition of the salts of the different samples. The relative proportions of salts and water do vary from place to place and with depth in the water column. Oceanographers define the ***salinity*** of seawater as the weight of the salts (in grams) in one kilogram (1000 grams) of seawater.

$$\text{Salinity} = \frac{\text{grams salt}}{\text{1000 grams seawater}}$$

The units of salinity are then parts per thousand (sometimes written *ppt*) or equivalently, per mil, and frequently shown by the symbol ‰.

Problem 4: You analyze a sample of seawater weighing 630 grams. You find that it contains 22 grams of salt and 608 grams of H_2O. What percent of the seawater is salt?

630 x 1.5873 = 1,000

22 x 1.5873 = 34.92

34.92 / 1,000

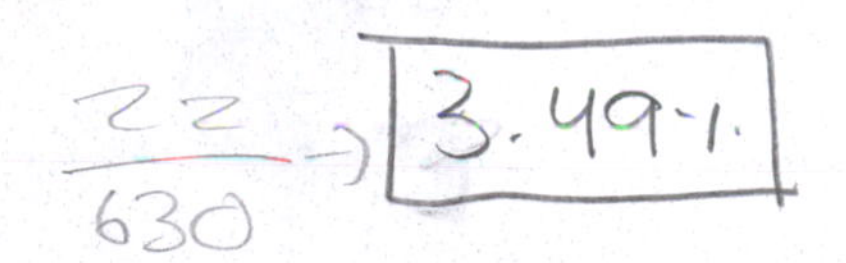

What is the salinity of the seawater in parts per thousand?

Freezing of seawater: As we add salt to water, the behavior of the solution changes in two ways pertinent to the present discussion. First, the more salt we add to the water the lower is the temperature at which the solution reaches its maximum density. Second, once we add salt to the water we can no longer define a single freezing point. The reason for this is explained below.

If we chill a salt solution it will reach a temperature at which ice crystals just begin to form. This temperature is known as the ***initial freezing temperature***. As the salinity increases the initial freezing temperature becomes lower. There is no room for salt to fit between the water molecules in ice crystals. As the ice forms, salt is excluded from the ice and the salinity of the remaining liquid water therefore increases. Thus, it is necessary to chill the remaining solution to still lower temperature in order to cause additional ice crystals to form. This combination of freezing plus cooling continues over an extremely extended temperature range. Finally, at a temperature of -21°C for a NaCl (sodium chloride) solution and somewhat lower for real seawater, the mixture can freeze into a solid mass of ice and salt crystals. (Students from cold, snowy climates may know that that is why salt is put on icy roads in the winter. As the salt interacts with the snow it causes at least some melting unless the temperatures are extremely cold.)

Two lines are drawn in Figure 8.3. The first shows the relationship between the salinity of a mixture of fresh water and seawater and the temperature at which it has maximum density. The second shows the relationship between salinity and initial freezing point. Effectively, there is no temperature

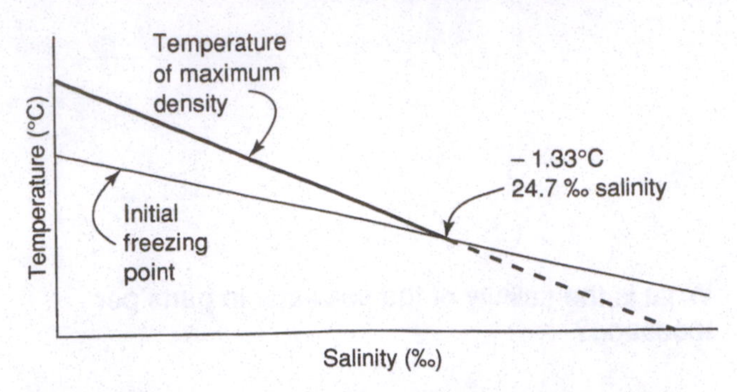

Figure 8.3. Variation of temperature of maximum density and of initial freezing point as a function of temperature. A solution with salinity greater than 24.7‰ does not have a temperature at which its density reaches a maximum. As such a solution is cooled, freezing begins before the temperature of maximum density is reached.

at which seawater of normal salinity reaches a maximum density, because when chilled it undergoes initial freezing at much higher temperature than the indicated temperature of maximum density. Once freezing starts, the residual water becomes saltier and accordingly the temperature of maximum density becomes even lower.

Temperature-density relationships of seawater: The density of seawater is determined by both the temperature and the salinity. (Pressure also affects the density of seawater slightly. Deep water is under great pressure as the result of the weight of the overlying water column. Consequently it becomes slightly compressed, i.e., slightly more dense. In this

Table 8.2. For use in Problem 5.

Sample	*Temperature*	*Salinity*	*Density*
1	16	36	1.026
2	0	34	1.027
3	25	35	1.025
4	24	35.5	1.024
5	14	37	1.028
6	4	37	1.029

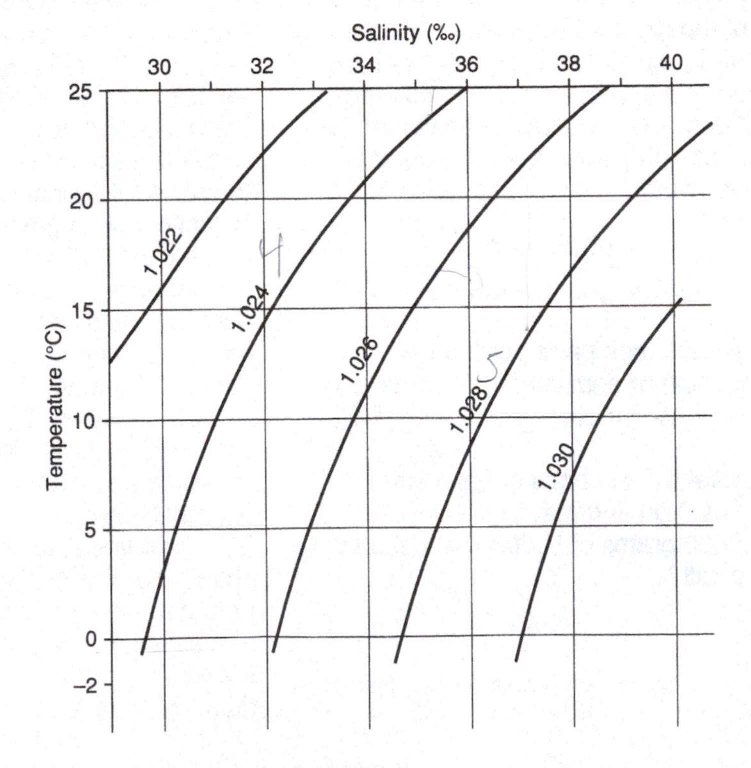

Figure 8.4. Relationship between temperature, salinity, and density for solutions made of water and sea salt. Curved lines, called ***isopycnals***, are lines of constant density. Water with a salinity of 34.6 per mil at a temperature of 0°C has a density of 1.028 gm/cm³. So does water with salinity of 38.0 per mil at a temperature of about 17°C, as does water with all combinations of temperature and salinity indicated by the curve labeled 1.028. (After G.L. Pickard, *Descriptive Physical Oceanography*, © 1963.)

exercise we ignore the effect of pressure on water density.) We can represent the densities of all conceivable seawater samples on the temperature-salinity (or T-S) diagram illustrated in Figure 8.4.

Problem 5: Using the temperature-salinity diagram in Figure 8.4, fill in the blank spaces on Table 8.2.

A *water mass* is a large volume of ocean water that has fairly uniform temperature and chemistry because it originates in a geographically restricted region as the result of a well defined process or set of processes. Oceanographers find temperature-salinity diagrams useful for understanding the relationships among water masses at different depths at the same location, as well as for tracing the evolution of water masses as they move through the oceans.

In Figure 8.5, the variations of temperature, salinity, and water density are plotted from the surface to the bottom at a site in the south Atlantic Ocean. In the case of this water column the temperature decreases ***monotonically*** with depth, while the salinity decreases, increases, and then decreases again. It is, in fact, not necessary that the temperature decrease with depth, but it is necessary that the density increases monotonically from the surface to the bottom.

The wiggles in T-S diagrams can be interpreted as indicating the mixing of different water masses. Imagine that at some place in the ocean there were three distinct water masses overlying

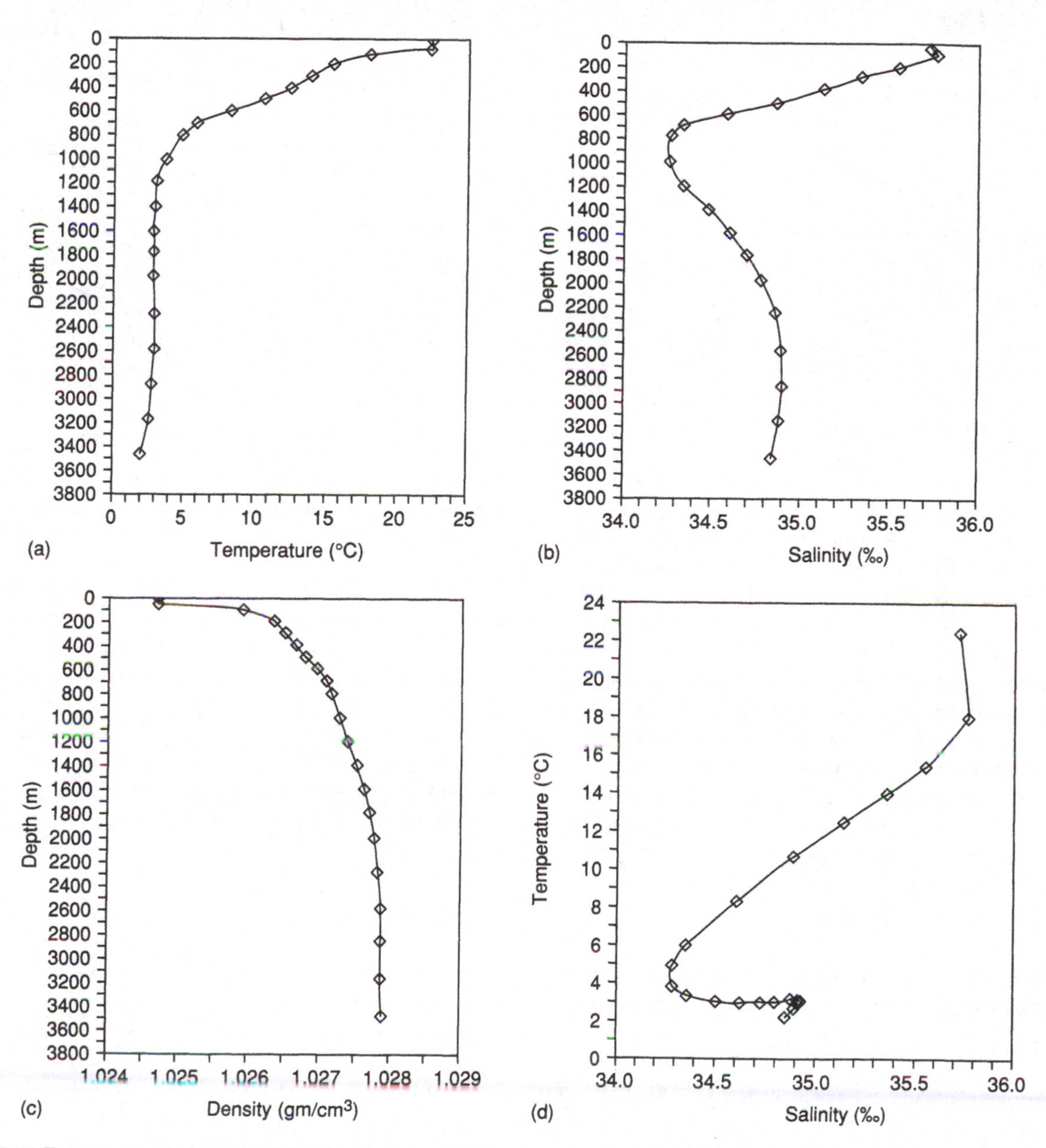

Figure 8.5. Temperature, salinity and water density data for a station in the south Atlantic Ocean at 33°S, 35°W. A, B, and C show the variation of temperature, salinity, and density, respectively, with depth. D is a temperature-salinity diagram.

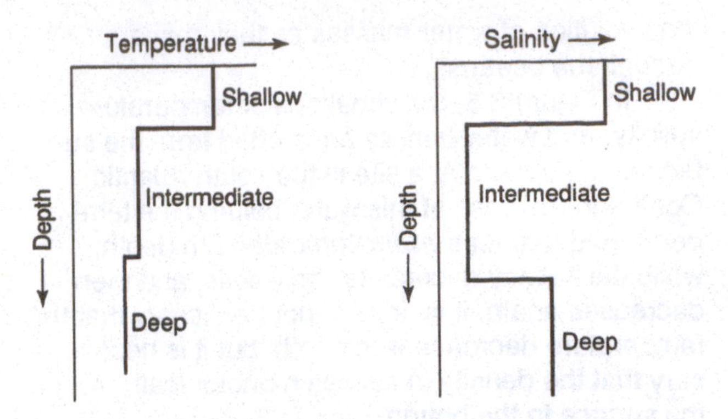

Figure 8.6. Variation with depth of temperature and salinity at a hypothetical location where the water column consists of three layers or water masses: *shallow, intermediate,* and *deep.* (Adapted from *Ocean Science,* Second Edition, fig. 15.7, by Keith Stowe. Copyright © 1983 by John Wiley and Sons.)

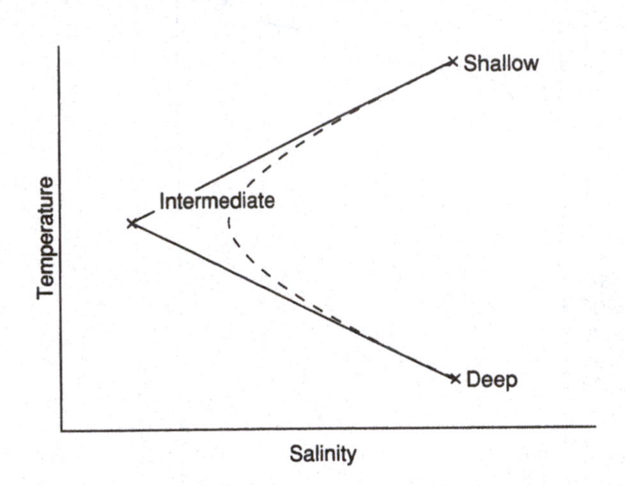

Figure 8.7. Temperature-salinity diagram generated from the hypothetical data in Figure 8.6. (Adapted from *Ocean Science,* Second Edition, fig. 15.7, by Keith Stowe. Copyright © 1983 by John Wiley and Sons.)

one another, each with its own characteristic values of temperature and salinity, as in Figure 8.6.

From the data in Figure 8.6 we can generate the T-S diagram shown in Figure 8.7. With time, there might be some mixing at the boundaries between the water masses and the distinct boundaries (represented by the sharp corner(s) on the T-S diagram) would be replaced by the more rounded corner(s) of the dashed curve.

Problem 6: Why is the T-S diagram in Figure 8.8 an impossible representation of a water column in the ocean?

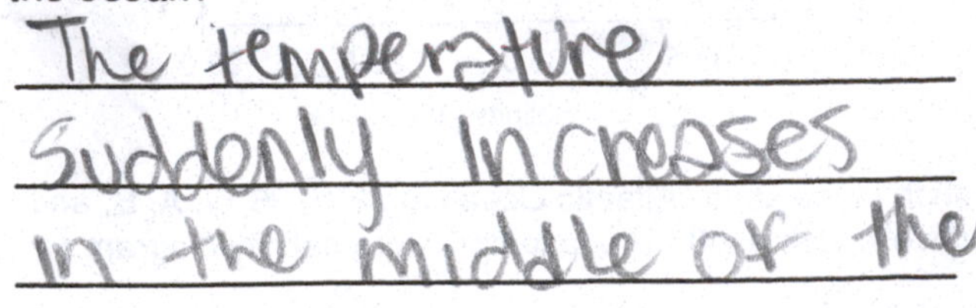

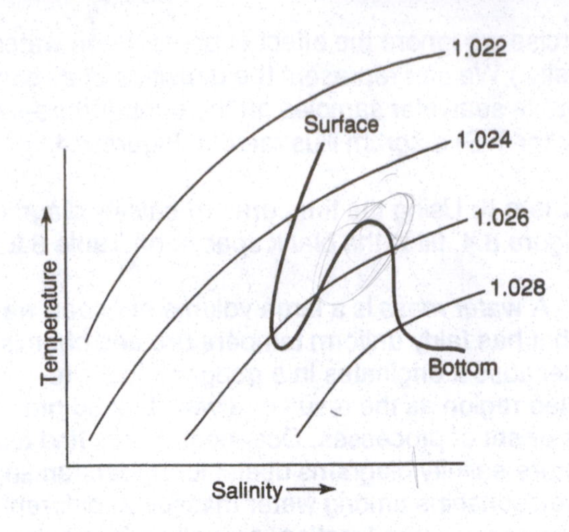

Figure 8.8. For use in Problem 6.

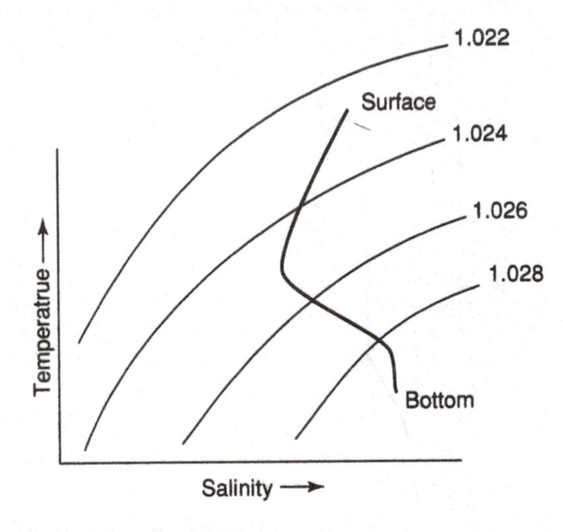

Figure 8.9. For use in Problem 7.

Surface & bottom

Problem 7: How many water masses are represented by the T-S diagram in Figure 8.9? How do you know?

Name: ______________________ Date: ____________ Section: ____________

Exercise 9
CORIOLIS EFFECT—OBJECTS MOVING ON A ROTATING EARTH

Purpose: In this exercise you will obtain a better understanding of the way in which objects move on a rotating earth, or the well-known ***Coriolis Effect***. The effect manifests itself by the apparent deflection in a clockwise direction (or to the right) of objects moving in the northern hemisphere and the apparent deflection in the counterclockwise direction (or to the left) of objects moving in the southern hemisphere.

Introduction: We have all been taught that once an object is set in motion it keeps moving at the same velocity in the same direction, provided that no new forces are brought to bear upon it. Imagine that we were on a planet on which the force of gravity somehow magically vanished for a few minutes. Imagine, further, that we were to shoot a cannonball northward, with the cannon aimed in the horizontal direction. The path the cannonball would take is shown in Figure 9.1. If the planet were (again, magically) not rotating the ball would follow the path labeled **a**. But if the planet is rotating about its axis as earth does, the cannon and the cannonball would both be moving from west to east. The actual path of the ball, labeled **b** in Figure 9.1, would be a combination of the northerly motion imparted by the cannon and the easterly motion imparted by the rotation of the planet.

Now let us look at a world in which the force of gravity functions normally. If we roll a ball across a smooth parking lot it appears to roll in a straight line, gradually slowing down as the force of friction converts its kinetic energy to heat. In fact, if the earth were not rotating about its axis and we observed the rolling ball from outer space, we would see that it would roll in a curved path; that is, the force of gravity would keep pulling the ball toward the center of the earth while inertia tried to keep it rolling in a straight line. The path that we would see the ball following would be defined by the intersection of the earth's surface with a plane passing through the earth's center (i.e., a great circle) as shown in Figure 9.2.

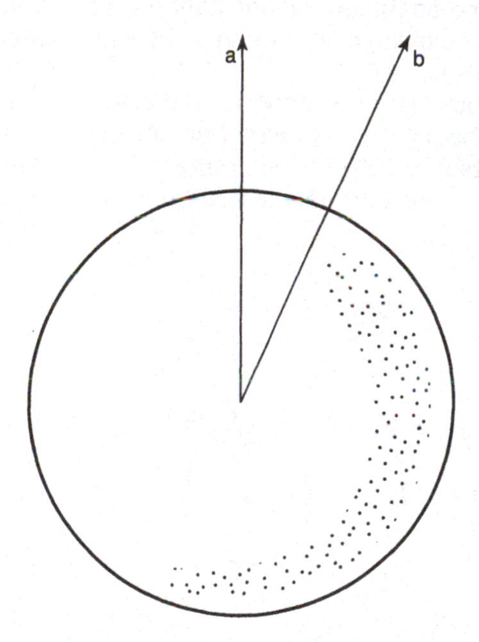

Figure 9.1. Moving objects travel in straight lines unless they are deflected by a force. This figure illustrates motion of a cannonball in a fantasy universe in which there is no gravitational force. The cannon is fired northward in the horizontal direction. Path **a** shows the trajectory of a cannonball fired on a nonrotating planet. Path **b** shows the trajectory when the planet is rotating from west to east, as the earth does.

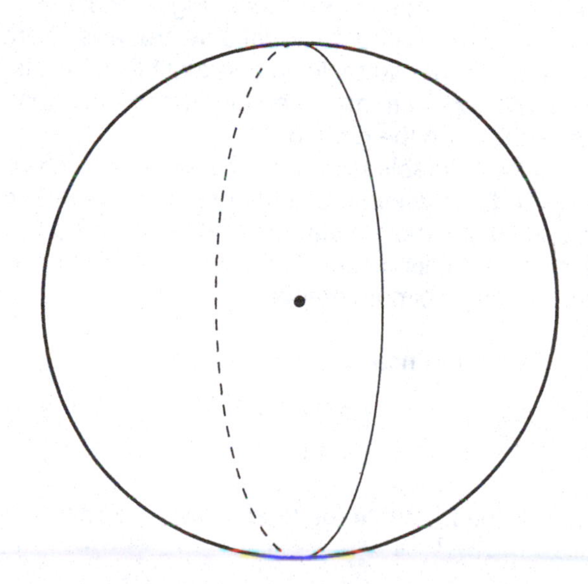

Figure 9.2. A ball rolling on the surface of the earth would roll in a circle if the surface were smooth and the earth were not rotating about its axis.

The real world, of course, spins about its axis of rotation, and that produces a striking result. Imagine, again, rolling the ball on the earth's surface. From our vantage point on the rotating earth, if we were able to look over large enough distances, we would see that in the northern hemisphere the trajectory of the ball would be diverted to the right of the straight-line (actually great circle) path while in the southern hemisphere it would be diverted to the left. It is as if some mysterious force is acting upon the ball, causing its trajectory to be diverted from its expected path. This "force" is sometimes called the Coriolis Force, but as we shall see, there really is no special force operating. It is preferable to speak of the Coriolis Effect.

The effect of which we have just spoken is the result of the fact that the objects are moving upon a rotating sphere, that they are constrained to move along the surface of a sphere (i.e., in a great circle rather than in a straight line as they would on a plane), and that we, the observers of the objects, are on the same rotating sphere. Had we been observing a moving object from a satellite in outer space, we would have seen that, except for being constrained by gravity to travel along the surface of the planet rather than shoot off into space, the path of the object continued, throughout its travel, in the same direction in which it started. This may sound confusing, but should become clearer as you proceed through the exercise.

Life on a rotating turntable: It is easiest to understand what is happening by looking first at a two-dimensional case. Imagine that you are living on a turntable like that of a record player. You have a toy cannon with you. If the turntable isn't turning and you fire the cannon outward as in Figure 9.3, the cannonball will appear to travel outward, away from the center of the turntable, in a straight line. This is how it will appear to an observer either on the turntable itself or on the ground.

If the turntable is rotating the situation is more complex. Let's assume that the diameter of the turntable is 50 cm and the cannon is located 6.37 (i.e., $20/\pi$) cm[1] from its center. Thus, the cannon is on a circle 40 cm in circumference.

$$\begin{aligned} \text{Circumference} &= 2 \times \pi \times \text{radius} \\ &= 2 \times \pi \times (20/\pi) \\ &= 40 \end{aligned}$$

Let the turntable rotate at 1 revolution per second in the clockwise direction. Before we shoot the cannon, *if you are sitting on the turntable,* it appears, in your frame of reference, that the cannon and the cannonball are not moving. But that is only an apparent lack of motion because you, the observer, are in the same rotating frame of reference as is the cannonball. (It is much the same as being on an airplane. The person across the aisle from you appears not to be moving, even though to an observer on the ground you may both appear to be moving at 500 miles an hour.) To the observer standing on the ground the cannon and the cannonball are moving clockwise in a circle at a speed of 40 cm/sec.

Now, fire the cannon and impart a radial velocity to the ball (i.e., outward from the center of the turntable) of 100 cm/sec. Your target is an enemy sitting on the turntable at a distance of 19.09 cm

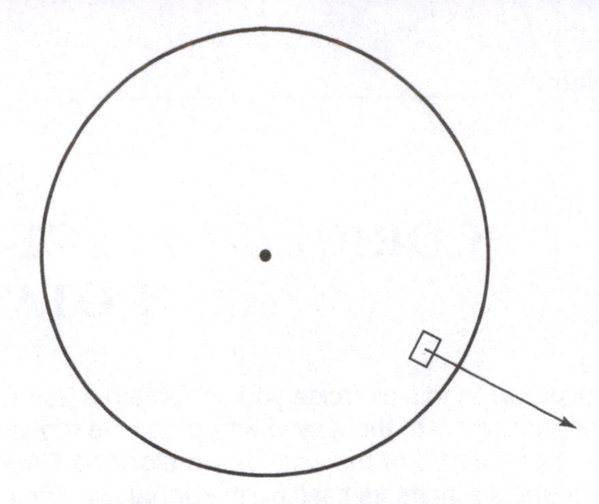

Figure 9.3. The trajectory of a cannonball fired from a *nonrotating* turntable, as seen by an observer on the turntable.

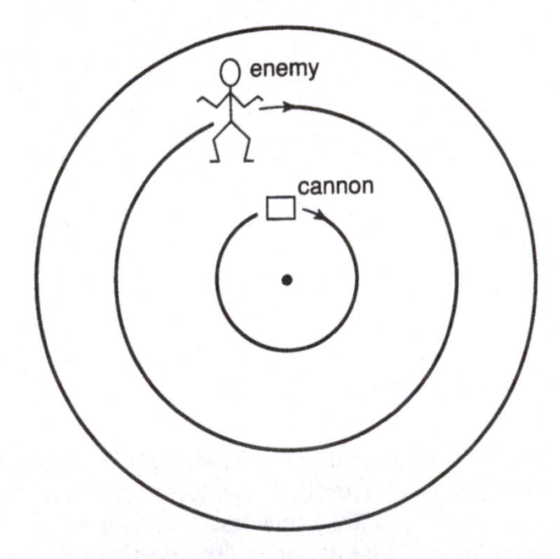

Figure 9.4. The turntable is rotating at 1 revolution per second. The cannon is rotating at 40 cm/second. At what speed (in cm/second) is the enemy, at a distance of $60/\pi$ cm from the center, rotating? (See Problem 1b.)

[1] π (the Greek letter *pi*) is the ratio of the circumference of a circle to its diameter. Its value is approximately 3.14.

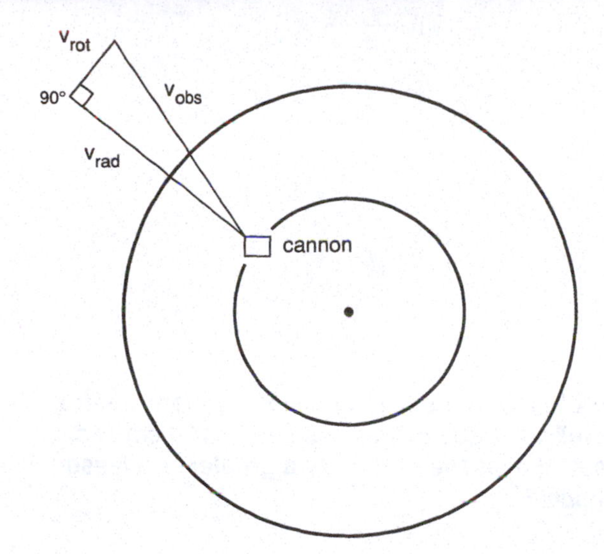

Figure 9.5. v_{rad} = radial velocity imparted by the cannon (100 cm/second). v_{rot} = velocity to the right imparted by rotation of cannon and cannonball prior to firing. v_{obs} = velocity seen by a stationary observer on the ground, and is the resultant of v_{rad} and v_{rot}.

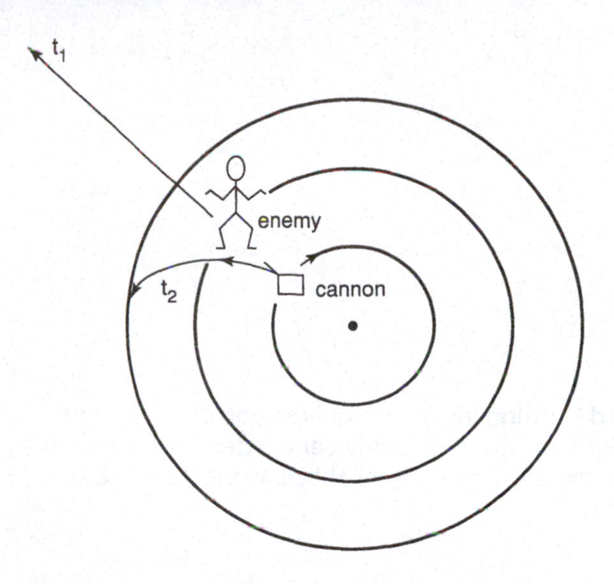

Figure 9.6. t_1 is the trajectory you thought the cannonball would follow because you didn't take the turntable rotation into account. t_2 is the trajectory of the cannonball as observed by you and your enemy, both on the rotating turntable.

(i.e., $60/\pi$) from the center (Figure 9.4).

To an observer on the ground, the cannonball will appear to travel in a straight line, as shown in Figure 9.5. But to you and your enemy, both sitting on the rotating turntable, the trajectory of the cannonball will appear to be diverted to the left by some mysterious force from the straight-line course you expected it to take. To your dismay and your enemy's delight the cannonball misses him, as shown in Figure 9.6.

Problem 1. Using the data in the turntable experiment described above, answer the following questions.

1a.) Remembering the ***Pythagorean Theorem*** for right triangles,

$$A^2 + B^2 = C^2$$

and applying it to the vectors shown in Figure 9.5, calculate V_{obs}, the speed of the cannonball as it is seen by an observer on the ground.

1b.) The cannon is rotating at a speed of 40 cm/sec. At what speed is the enemy rotating?

1c.) Although it is not a completely correct assumption, let us assume that you can neglect all but the radial component of motion in this part of the problem. How long does it take for the cannonball to travel the distance from the circle on which it lies to the circle on which the enemy is sitting?

1d.) During the length of time you calculated in Problem 1c, how many centimeters to the right (i.e., in the clockwise direction) does the cannonball move?

1e.) During the same length of time, how many centimeters to the right does the enemy move?

1f.) Given your answers to Problems 1d and 1e, by how many centimeters does your cannonball miss your enemy?

Problem 2: If the cannonball in Problem 1 were travelling at 200 cm/sec instead of 100 cm/sec, would it miss the enemy by a greater or a lesser amount?

____

Explain.

____

____

____

____

____

____

____

____

Life on a rotating cylinder: Now let us imagine that we live not on a turntable, but on a cylinder that rotates about its axis of rotation, as seen in Figure 9.7.

As on the turntable, you shoot a cannonball from your cannon (point **A**) directly north toward your enemy (point **B**).

Problem 3: If the radius of the cylinder is 8000 miles and the cylinder makes one rotation each day, calculate the rate of rotation of point **A** (the cannon) in miles per hour.

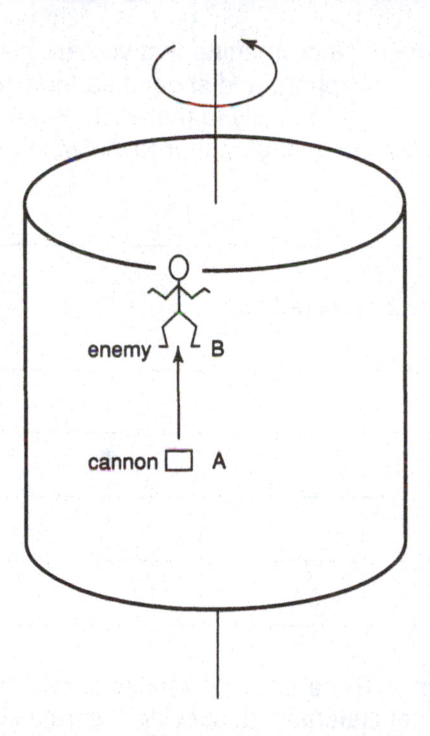

Figure 9.7. A hypothetical planet, shaped like a cylinder and rotating about an axis. A cannon located at point **A** shoots a cannonball northward toward a target at point **B**.

Compare that to the rate of rotation of point **B** (the target).

Using the above answers, will a cannonball shot due northward from point **A** to point **B** be deflected to the left of the target, not be deflected at all, or be deflected to the right of the target? Explain your reasoning carefully.

____

____

____

____

____

____

Life on the real world: Having examined how objects move on a rotating flat turntable and a rotating cylinder, we are better able to consider what happens to an object moving on a spherical earth. If the earth were not rotating, the motion of an object would still be affected by the force of gravity. So it would not travel in a straight line. Instead it would travel in a great circle (or in an elliptical orbit, if, as an artificial satellite, it is free to leave the surface of the earth). The earth, of course, does rotate. To an observer in space, an object set in motion on the earth appears to follow a trajectory determined by the initial force imparted to it, modified by the earth's gravitational pull. However, to an observer on the earth, as to the observer on the rotating, two-dimensional turntable, a moving object appears to be deflected by some mysterious force.

Problem 4: If you were sitting in outer space above the North Pole, would the earth appear to be rotating clockwise or counterclockwise?

____

Explain your reasoning.

____

____

____

____

____

Problem 5: Would you expect the apparent deflection of an object to be greater at the equator or at the pole?

____

Explain your reasoning using the logic that was developed in the examples of the rotating turntable and the rotating cylinder.

____

Problem 6a: Imagine that you are in the southern hemisphere and shoot a cannon northward, at a target directly to the north (but still in the southern hemisphere). Would the cannonball land to the east or to the west of the target?

Explain your reasoning.

Problem 6b: Now, imagine that you are in the northern hemisphere and shoot a cannon northward, at a target directly to the north. Would the cannonball land to the east or to the west of the target?

Explain your reasoning.

Problem 7: Generalize the results of Problem 6 with a brief statement describing the trajectories of objects moving on the real earth.

Name: ______________________________ Date: ______________ Section: _____________

Exercise 10
ATMOSPHERIC CIRCULATION AND SURFACE OCEAN CURRENTS

Purpose: In this exercise you will become familiar with the major features of the earth's atmospheric circulation and the major surface currents of the oceans. You will examine the relationship between the two, and see the way in which they are affected by the earth's rotation (the Coriolis Effect discussed in Exercise 9).

The surface currents of the oceans are largely independent of the circulation of the deeper waters. The boundary between the surface circulation and the deep circulation can be taken to be the ***thermocline*** (discussed in Exercise 8). The main driving force for surface circulation is the wind, which causes the water to move as it blows across the water. Thus, the circulation of the surface waters is sometimes referred to as the ***wind-driven circulation***.

Atmospheric circulation: The intensity of solar energy (expressed as amount of energy per square centimeter) available to heat the surface of the earth is greater at the equator than at the poles. The three reasons for this can be seen in Figure 10.1. First, because of the spherical shape of the earth, the same amount of solar energy is spread over a larger area at high latitudes than at low. Second, at high latitudes a ray of solar energy travels through a greater thickness of atmosphere and therefore has greater chance to be reflected or scattered back to outer space before reaching the surface. Third, the sun's rays strike the earth's surface at lower angles at high latitudes, and are more likely to be reflected back into space.

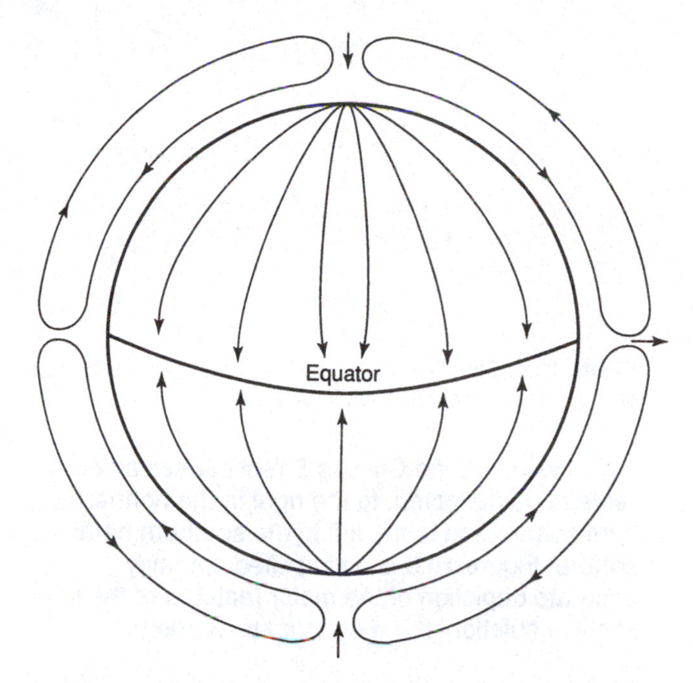

Figure 10.2. If the earth did not rotate about its axis, the atmosphere might circulate as shown. In this sketch, each hemisphere is characterized by three convection cells.

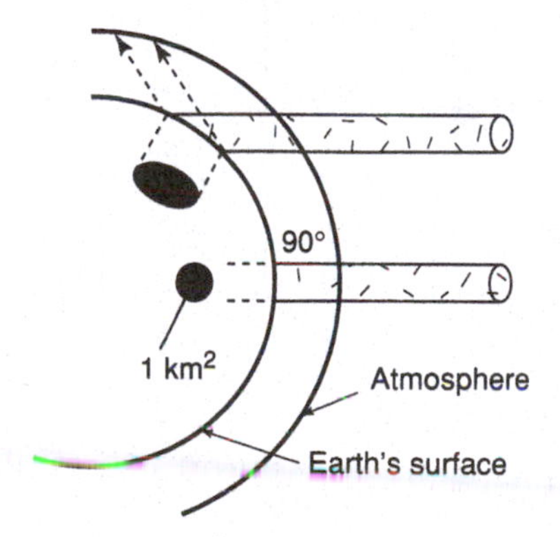

Figure 10.1. The intensity of solar radiation available to warm the surface of the earth is much greater at the equator than at the poles, as the result of the shape of the earth.

The earth not only absorbs heat from the sun. It radiates heat back to outer space. At the equator the earth absorbs more heat than it radiates, while at high latitudes it radiates more than it absorbs. This would seem to suggest that the tropics ought to be warming while the poles are cooling. But that is *not* the case. Rather, the surplus energy of the equatorial regions is transported toward the poles by means of the atmosphere and the oceans, making up the energy deficit at high latitudes. Another way of looking at this is that the atmosphere and the oceans constitute a great engine that is put in motion by temperature differences at high and low latitudes.

If the earth were not rotating, we might expect large convection cells to develop, with warm air rising and cool air sinking. The circulation of the atmosphere might appear as in Figure 10.2.

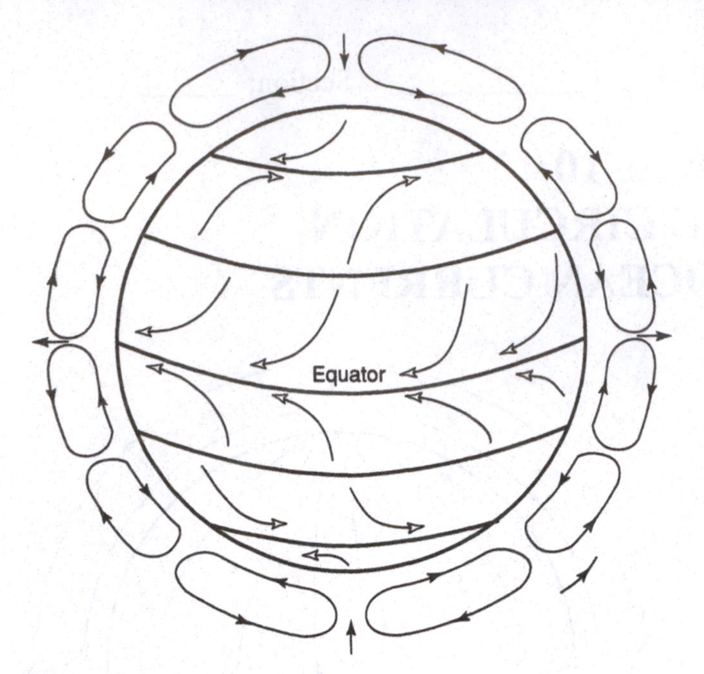

Figure 10.3. Simplified sketch of the large scale features of the earth's atmospheric circulation.

However, the Coriolis Effect causes air currents to be deflected, to the right in the northern hemisphere and to the left in the southern hemisphere. Figure 10.3 is a simplified but fairly accurate depiction of the major features of the large scale circulation of the earth's atmosphere.

Problem 1: Remembering that warm air can contain more water vapor than cold air, use Figure 10.3 to explain why satellite photographs of the earth often show the equatorial region to be cloudy.

The equator tends to be warmer so the warm air makes more clouds.

Problem 2: Again using Figure 10.3, explain why many of the great deserts of the world are at (approximately) 30 degrees north or south of the equator.

The equator is hit by the sun the most intensly, so it would make sense that the deserts are located near the equator.

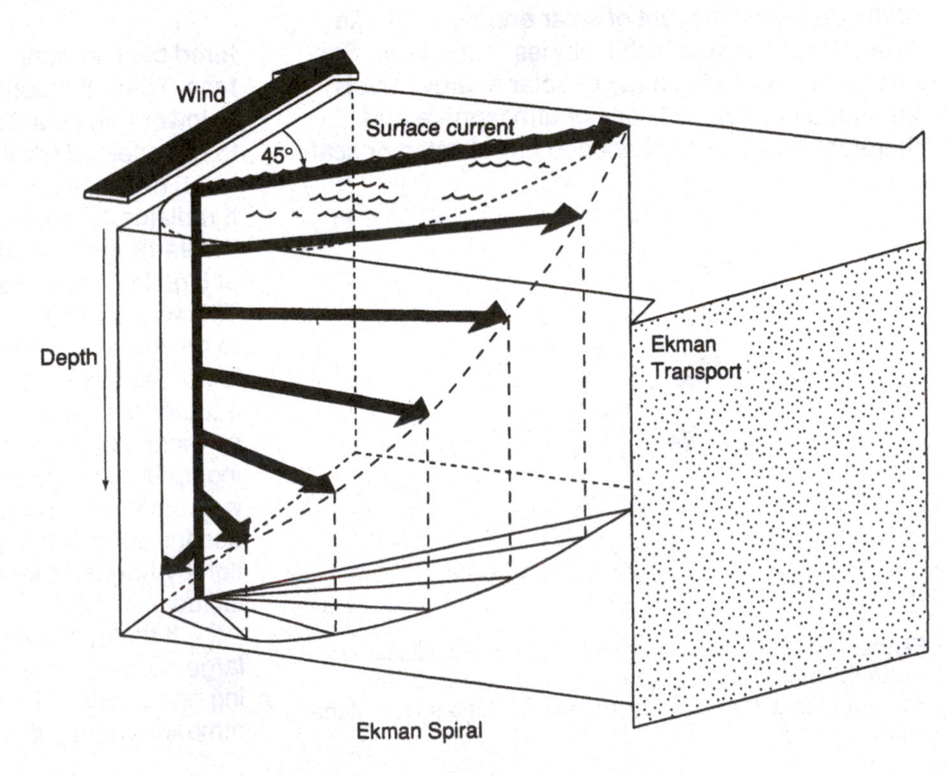

Figure 10.4. The *surface wind drift*, or wind driven current, moves (in the Northern Hemisphere) in a direction to the right of the direction in which the wind is blowing. Imagine the water column as consisting of a very large number of layers stacked on top of one another. Each layer is pushed by the overlying layer and is held back by frictional drag with the underlying layer. Each arrow represents the velocity of a layer. The length of each arrow is proportional to the velocity of the layer and the arrow points in the direction in which the current is moving. Each layer of water moves to the right of the overlying layer, but moves more slowly than the overlying layer because frictional drag with its underlying layer slows it down.

The Ekman Spiral: As the wind blows across the water it sets the water in motion. However, the water does not move exactly in the same direction as the wind. This is a result of the Coriolis effect. The ***Ekman Spiral*** is an idealized representation of the way in which water moves as the wind blows across it. Early observations of the fact that surface currents in the Northern Hemisphere flowed in a direction to the right of the wind were made during the 1890s from the Norwegian ship *Fram* under the direction of Fritdjof Nansen. The *Fram* was purposely frozen into pack ice in the Arctic Ocean and drifted for about three years. During that time it was noted that the ship drifted consistently to the right of the direction in which the wind was blowing. The effect was explained mathematically by the Swedish oceanographer V. W. Ekman, and the spiral-like motion of the water he described bears his name.

The mathematical treatment of the Ekman spiral is too complex to deal with in this laboratory manual, but we can quite readily visualize qualitatively what is occurring. Imagine a column of water as a stack of extremely thin sheets of water piled upon each other, each sheet moving horizontally, as shown in Figure 10.4, in which the direction of the arrow indicates the direction of water motion and the length of the arrow is proportional to the velocity. The wind will set the uppermost layer in motion, but as soon as it begins to move the Coriolis effect will cause its motion to deviate from that of the wind by about 45 degrees (to the right in the northern hemisphere). Friction between the first layer and the second layer will set the second layer in motion, and it will move a little to the right of the first layer. The second layer will be held back a little by friction with the third layer, which will in turn be set into motion by the second layer, and so on and so on. The result will be that each layer of water will move a little more slowly than the layer above it and will move slightly to the right of the layer above it (or to the left in the southern hemisphere). Theoretical calculations indicate that at some depth the water will be moving in exactly the opposite direction to the wind (although not very fast) and that the net motion of the water column is at right angles to the wind.

The net motion of water to the right (or left in the southern hemisphere) of the wind as described above is known as the ***Ekman transport***. The Ekman transport is quite evident in the way it causes nutrient-rich waters from depths of a few hundred meters to upwell in some coastal zones. (This upwelling, when it occurs, can significantly increase biological productivity.) As shown in Figure 10.5, if the wind is blowing parallel to the shore the net movement of the surface water will be perpendicular to and either toward the shore (onshore) or away from the shore (offshore), depending on the hemisphere and on which of the two possible directions parallel to the shore the wind is blowing. If the motion of surface and near-surface water is offshore, deeper water will have to flow in to take the place of the water that is moving out to sea. As explained later in this exercise, the ***upwelling*** that develops will enhance biologic productivity in the area. If the motion of the surface and near-surface water is toward the shore, that water will have to go somewhere when it meets the land, typically downward into the deeper layers of the ocean, and the pattern that will develop will be one of ***downwelling***.

Problem 3: Does Figure 10.5 apply to the northern hemisphere or the southern hemisphere?

Northern

Explain your answer.

The wind is pushing the water away from the coast

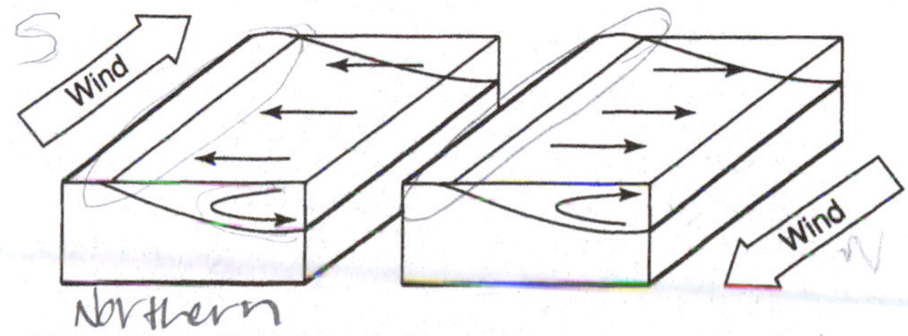

Figure 10.5. This figure shows the response of the currents when the wind blows parallel to the coast. But in which hemisphere? See Problem 3.

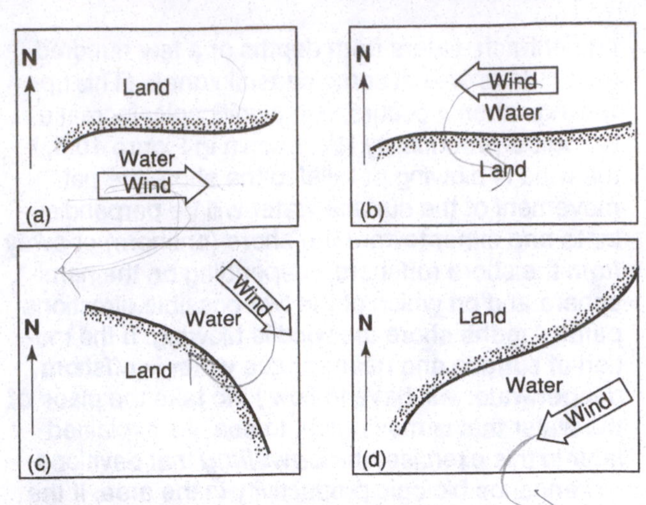

Figure 10.6. Hypothetical maps of wind blowing along coastlines, for use in Problem 4.

Problem 4: In Figure 10.6 are maps of four coastlines in the southern hemisphere. Wind direction is shown in each case. Indicate on each map whether the coastline will experience upwelling or downwelling.

The Coriolis effect is also very important on a large scale in governing ocean-wide movements of water. Figure 10.7 is a map of the major surface currents of the world. Note that each of the major ocean basins (North Atlantic, South Atlantic, North Pacific, South Pacific, and South Indian) is dominated by a large, more or less circular current, and that those currents flow clockwise in the northern hemisphere and counterclockwise in the southern hemisphere. Each of these circular currents is called a ***subtropical gyre***. (Some parts of some of the currents have familiar names. For example, the western part of the North Atlantic subtropical gyre is called the ***Gulf Stream***.) These gyres are ***geostrophic*** currents, meaning that their patterns are determined by the rotation of the earth.

The geostrophic currents are generated as described below. Imagine an idealized ocean basin in the northern hemisphere as shown in Figure 10.8. In the trade wind zone the winds are blowing from northeast to southwest, and the resulting Ekman transport causes water to move more or less toward the northwest. In the mid-latitude zone winds are variable, but are most often from the southwest. (See Figure 10.3.) In that case the Ekman transport causes water to move toward the southeast. The result is that water tends to pile up, or form a bulge or a hill in the central portion of the ocean basin.

Imagine the motion of a particle[1] of water on the surface of a bulge in the middle of the ocean. Gravity will cause the particle to move downhill. But as the particle begins to move downhill its path will

[1] Think of a particle of water as a very small volume of water, perhaps the size of a few drops, that moves as a unit, at least for a time.

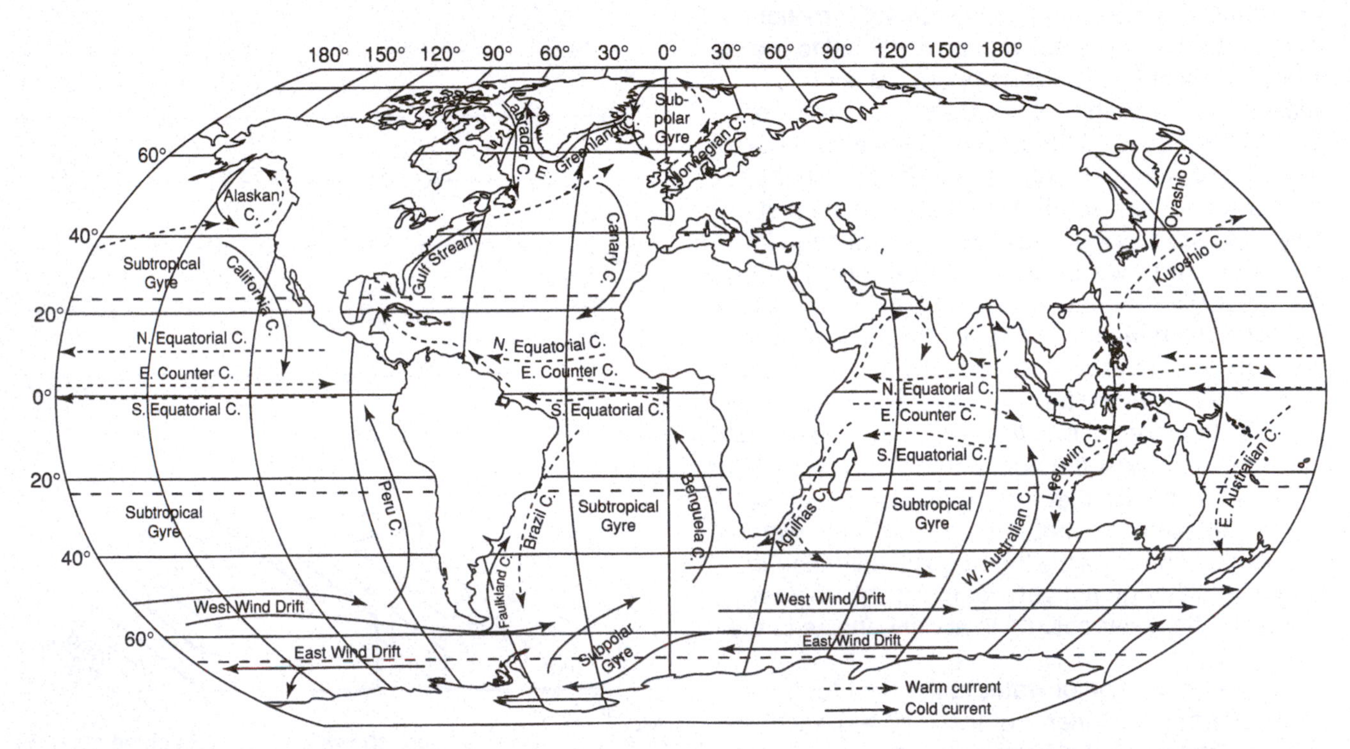

Figure 10.7 The major surface currents of the world in February and March. (Adapted from H.U. Sverdrup, M.W. Johnson, and R.H. Fleming, *The Oceans*, © Prentice-Hall, Inc., 1942.)

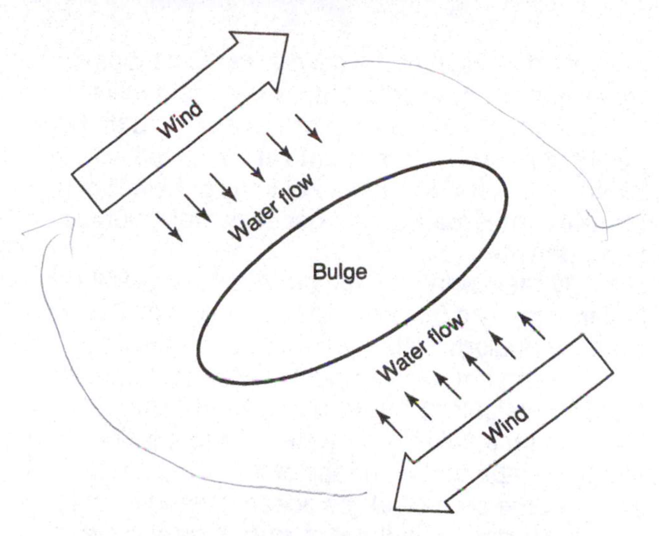

Figure 10.8. Generation of the bulge which occurs in the center of each of the major oceanic basins, and about which the currents of each subtropical gyre flow. This diagram is for the northern hemisphere.

be deflected to the right according to the Coriolis effect. If there were no friction between particles of water as they moved through the ocean, the moving particle of water would be deflected to the right until the magnitude of the Coriolis effect causing it to move uphill just counterbalanced the force of gravity pulling it downhill. The particle would then travel round and round the bulge of water at a constant elevation. The current that resulted would be considered an ideal geostrophic current. In the real ocean there is of course friction between a particle of water and neighboring particles so the particle of water would gradually slow if new energy were not provided. As the particle slowed, the Coriolis effect would no longer exactly counterbalance the force of gravity, and the particle would begin to move down the bulge again under the force of gravity, picking up speed as it went. Because of friction, the actual current deviates somewhat from the ideal geostrophic current in that, rather than running round and round the bulge, it would slowly spiral down the bulge as seen in Figure 10.9.

The height of the bulge in the middle of the

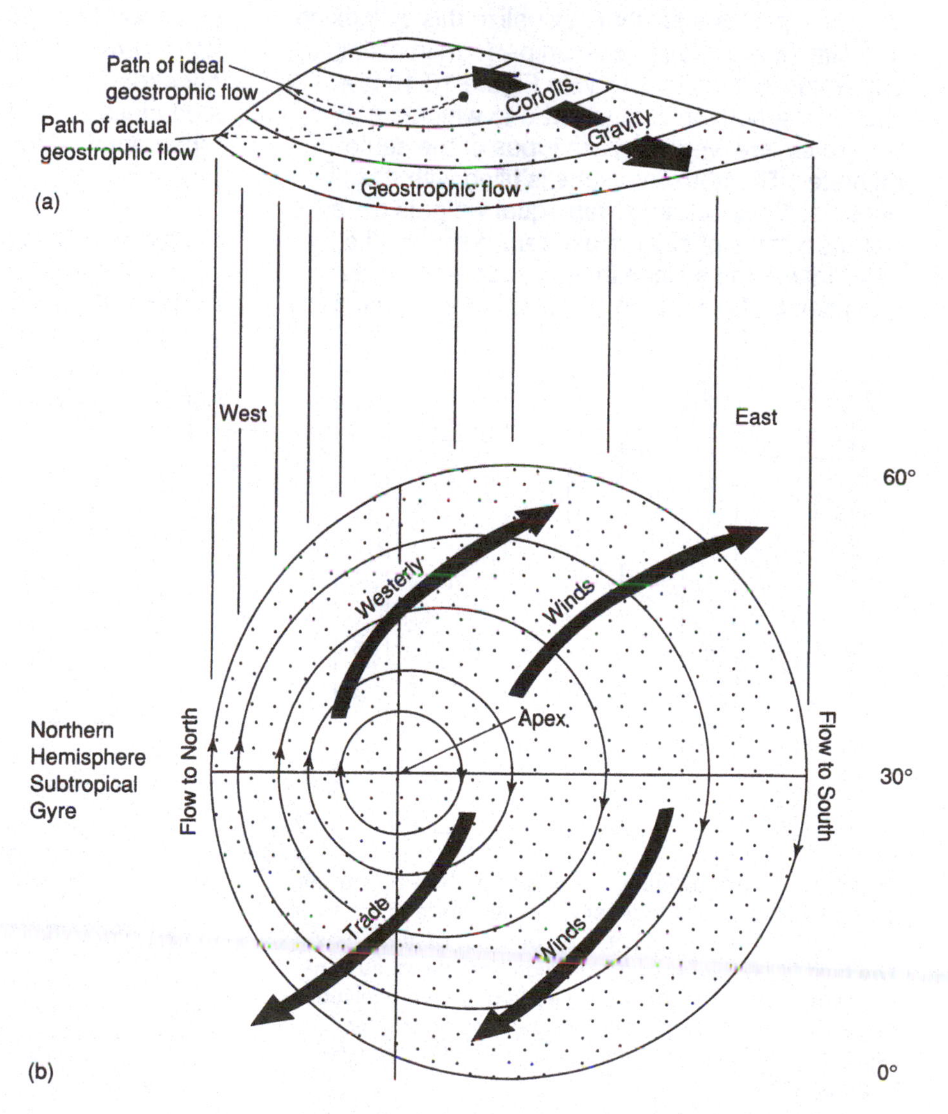

Figure 10.9. A cross-sectional view, A, and a map view, B, of the bulge in the surface of the ocean in an idealized, mid-latitude, Northern Hemisphere ocean. As a particle of water flows downward off the mound under the force of gravity its direction deviates to the right (in the Northern Hemisphere) by the Coriolis effect. The path of the particle is shown by the dashed arrow. As the particle travels downward off the mound, it swings more to the right, until at some point the gravitational force and the Coriolis force just balance, causing the particle, in the ideal case, to circle round and round the bulge. In the real case, shown by the dotted arrow, friction with surrounding water slows the particle down, and it gradually spirals off the mound.

ocean is not very great—perhaps a couple of meters or so. In recent years it has become possible to measure the height of the bulge directly, using modern measuring instruments on artificial satellites. But it is instructive to consider the way the topography of the sea surface was determined for many years prior to the development of modern, remote-sensing methods. It is possible to estimate the shape of the surface of the ocean indirectly, using temperature and salinity measurements. In fact, it is the usefulness of temperature and salinity measurements for calculating sea surface topography that has been one of the main motivating forces for making large numbers of such estimates.

We can imagine that there exists in the oceans a depth below which the motion of water generated by particles moving on the bulge created by the winds has died out. We can take that depth (or any depth greater than that) as an arbitrarily chosen reference level or *depth of no horizontal motion*. We can then determine the heights of columns of water of equal weights above the reference level at different locations.

It is perhaps easier to visualize this by looking at a simple experiment we can perform in a simple apparatus in the laboratory. In Figure 10.10 is a sketch of a horizontal glass tube to which are attached several vertical glass tubes of the same diameter. The horizontal tube is filled with a fairly dense salt solution and then equal weights of seawater samples of several different densities (i.e. different salinities since in the experiment everything can be taken to be at the same temperature)

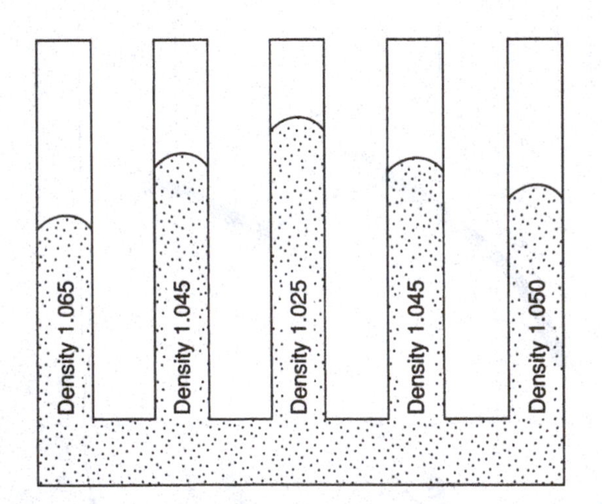

Figure 10.10. An experimental setup for showing the relationship between water density and sea surface topography. The device could be built out of glass tubing. The fluid filling the horizontal portion has the same density everywhere, but the fluid in each of the vertical tubes has a different density. The fluid columns of greater density will stand lower than the columns filled with fluid of lower density.

are added to each of the several vertical tubes. If we fill each of the vertical tubes very carefully to prevent any mixing of solutions, we will find that the height of the seawater column in each tube will be different. Shorter seawater columns of higher density will counterbalance longer seawater columns of lower density.

By measuring the temperature and salinity of water at many different places in the ocean it is possible, by analogy with Figure 10.10, to calculate the density, and hence the height of water columns of identical weight[2] at different locations. Then the heights of the columns of water of equal weight define the shape of the surface of the ocean. The shape of the surface of the ocean defined in this way is referred to as the ***dynamic topography***.

If we know the shape of the surface of the ocean we can calculate the speed of geostrophic currents generated as water is pulled by gravity downward from topographic high spots to lower spots. Not surprisingly, the steeper the slope of the water surface (i.e., the closer together the contours on a topographic map of the surface) the faster we can expect the geostrophic currents to flow. Also, not surprisingly, the higher the latitude the faster the geostrophic current we can expect for any given steepness of the dynamic topography. The dynamic topography of the north Atlantic Ocean is shown in Figure 10.11.

Problem 5: Draw on Figure 10.11 *in pencil* the region of the map in which the surface currents flow most swiftly, and indicate using arrows the direction in which they flow.

Briefly explain why you chose the region and current direction you did.

We can assume that's
the fastest because
the lines are closer
together

[2] Strictly speaking, it is not the weight but the pressure at the base of each column that is the same. The pressures are determined by the densities and the heights of the columns. If both the pressure at the base of each column and the cross-sectional area of each column are the same, the weights of all the columns must be equal.

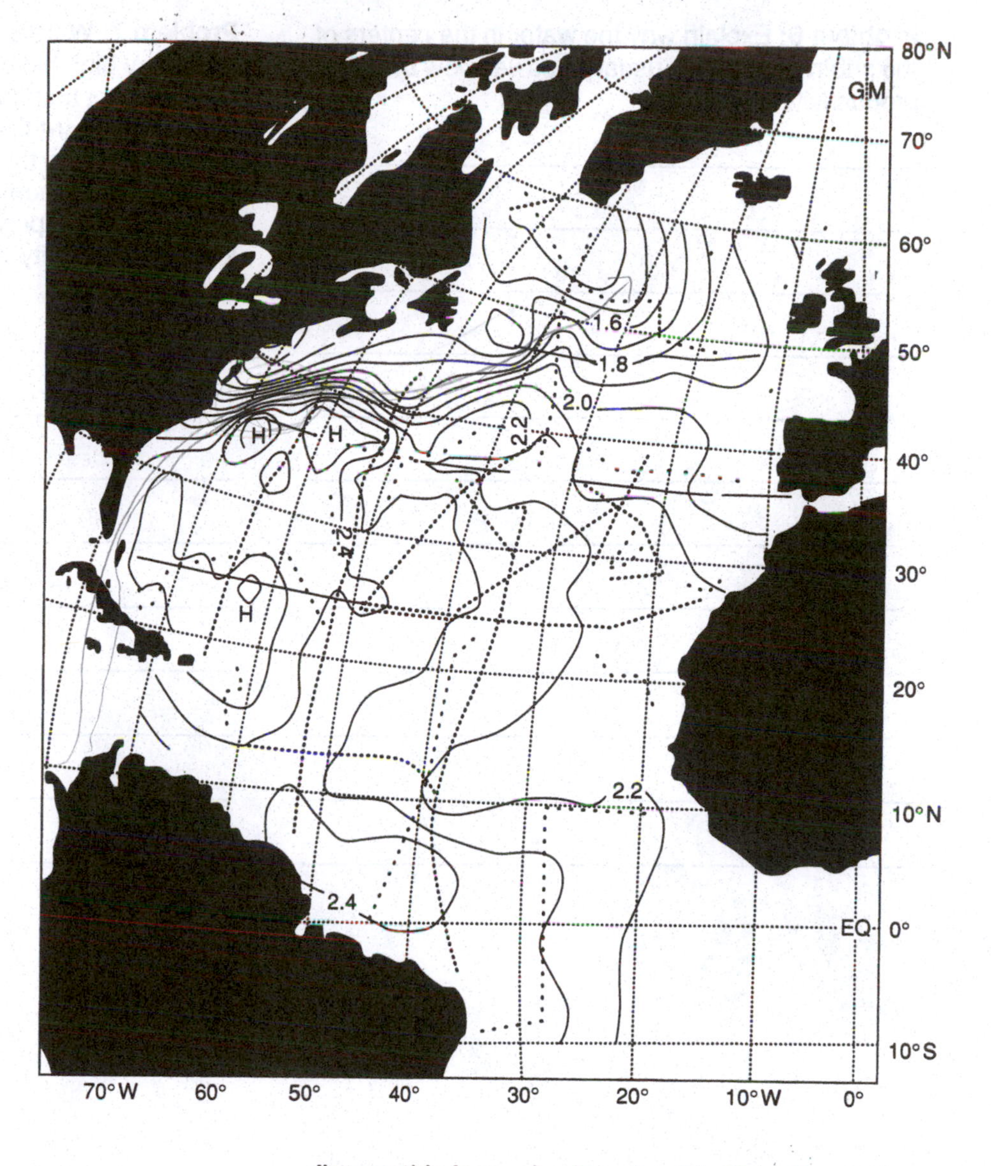

Figure 10.11. Dynamic topography of the north Atlantic Ocean. Numbers are elevations (in meters) of the water surface above an arbitrary level. (Reprinted from *Progress in Oceanography*, Volume 27, I. Fukumori, F. Martel and C. Wunsch, The hydrography of the North Atlantic in the early 1980s. An Atlas, Copyright 1991, with kind permission from Elsevier Science Ltd., The Boulevard, Langford Lane, Kidlington OX5 1GB, UK.)

____

____

____

Ocean circulation and biological productivity: In the center of the bulge in the middle of a subtropical gyre the thickness of the column of low density surface water overlying higher density deeper water is greater than on the outside of the bulge. This is because surface waters have piled up in the center of the bulge.

Plant growth, or ***primary production***, provides the basis of nutritional support for all animals. When we speak of ***biological productivity***, we mean the rate of plant growth in a region. Plant growth requires nutrients, especially nitrate and phosphate, and the availability of those nutrients limits the biological productivity of a region. We will discuss this in much more detail in Exercise 16, but the discussion below provides sufficient information to answer the questions of this exercise.

As plants grow, they remove nutrients from the water and incorporate them in their tissues. There is a general downward movement of the nutrients as the plants die and sink, or as animals eat the plants and then either die and sink, or as material excreted by the animals sinks. Gradually, bacterial decay of sinking biological material releases the nutrients back into solution. But typically, most of the nutrients are released back into the water at depths below which light penetrates, and hence at depths where plants cannot make use of them. This ***biological pumping*** of nutrients downward results in reduced concentrations of nutrients in surface waters, and hence in reduced biological productivity. Biological productivity is usually highest where upwelling of nutrient-rich water from depths of a few hundred meters brings nutrients back to the surface where plants can make use of them.

Problem 6: Explain why the water in the centers of the subtropical gyres usually has very low biological productivity.

In the center of the gyre the water is being pulled and the downwelling is taking all the nutrient rich water, so organisms aren't sustainable there.

Problem 7: Waters in the equatorial zone usually have very high biological productivity. Keeping in mind the fact that the trade winds blow from the northeast toward the equator in the northern hemisphere and from the southeast toward the equator in the southern hemisphere, and taking into account the Ekman transport of the water, explain why biological productivity is so high in the equatorial region.

Since the water is being pulled upward near the equator. There's a lot of nutrient rich water there, that allow for a higher biological productivity.

Name: ______________________________ Date: ______________ Section: ____________

Exercise 11
WAVES

Purpose: In this exercise you will get experience dealing with the nature of several properties of waves, including orbital motion, wave dispersion, and wave interference, and you will learn how to estimate the heights of waves generated by storms.

The Motion of Water in Waves: We know from watching objects floating downstream in rivers that floating objects move along with the water in which they float. It is convenient to think of any body of water as being made up of *parcels of water*, or masses or blobs of water with imaginary boundaries of any shape one cares to choose. We can think of a floating object (say, a log) in the river as being pushed along by the parcels of water surrounding it as they move downstream.

Now think of the appearance of waves traveling across the surface of the water in a lake or ocean. If the water is relatively deep and we watch a log floating on the surface we may see something different than we saw when we watched the log floating downstream in the river. The waves may appear to travel across the water surface, but the log does not travel along with them. At first glance it appears to bob up and down upon the waves. If we look more carefully we see that, with each passing wave, the log oscillates back and forth a bit in addition to moving up and down. (Still closer observation indicates that it creeps forward, i.e. in the direction of wave travel, a very little bit with each passing wave, but we will ignore that forward creep in this exercise.)

The primary motion of our log, or any other floating object, as the waves pass, is circular motion in a vertical plane. As indicated above, that circular motion is created by similar circular motions of the parcels of water surrounding the log. This motion is referred to as ***orbital motion***.

It is important to remember that the evidence we see when we watch the floating log tells us that the motion of the water itself is very different from the motion of the waves. If you like to swim in large ocean waves, you may have felt this circular motion yourself as you floated out beyond the breaker zone at the beach. (This motion is hard to feel in the smaller waves you are likely to encounter in most lakes and even many ocean beaches.) The orbital motion of a parcel of water at the lake or sea surface as waves pass is shown in Figure 11.1. Note that a parcel of water at the surface always stays essentially at the surface as the wave passes.

Problem 1: Describe the orbital motion you would feel if you were floating in the waves out beyond the breaker zone at the beach. Specifically, what motion would you feel at the very crest of the wave?

____

____

____

____

____

____

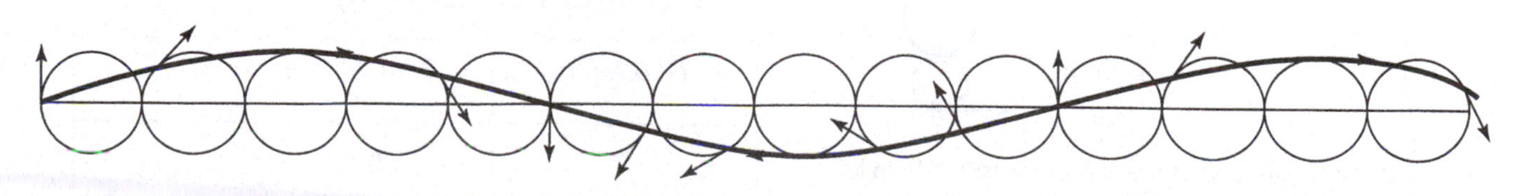

Figure 11.1. Cross section of a wave illustrating the orbital motion of parcels of water at the surface of a wave that is progressing from left to right. The shape of the surface at one particular instant is shown by the heavy curve. Each circle depicts the motion of a parcel of water at the surface of the ocean during the passage of a single wave. The base of each arrow indicates the location in the orbit for selected parcels of water when the wave is in the position shown. The direction of each arrow indicates the direction in which that parcel is moving when the wave is in the position shown. For example, the left-most parcel of water is moving straight upward.

What motion would you feel in the bottom of the trough?

____

____

____

____

Describe the motion you would feel as the wave advanced and carried you from the trough to the crest.

____

____

____

____

____

____

____

____

As we look at the motions of parcels of water beneath the surface, we find that the deeper we go the smaller the diameter of the orbit becomes. This is shown in Figure 11.2.

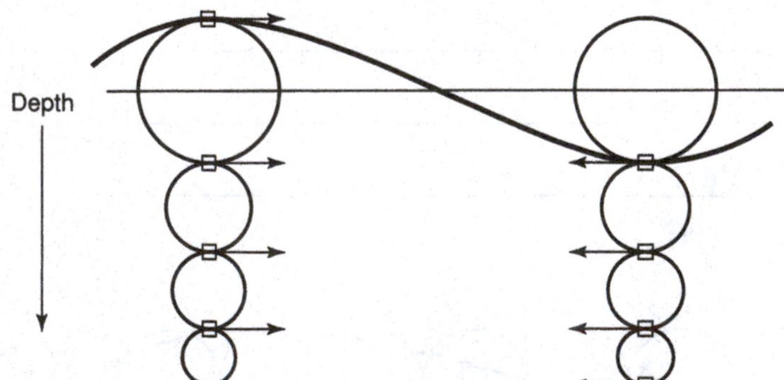

Figure 11.2. Cross section of a wave progressing from left to right, showing the decrease in orbital motion with increasing depth. Circles at the left show orbits of several parcels of water under the crest of a wave. Each parcel is indicated by a small square. The arrows show the direction of motion of those parcels. The circles at the right show orbits of several parcels of water under the trough of a wave.

Note that the individual parcels of water do not mix with one another. That is, underneath any given point at the surface, all the parcels of water move up, down, or sideways more or less together. The orbits get progressively smaller with increasing depth. As a rough rule of thumb, we can say that the orbital diameter decreases by approximately one half every time we increase the depth by (1/9) x L, where L is the wavelength.

Problem 2: Imagine a wave with a wavelength of 180 cm (approximately 6 feet). The orbital diameter at the surface is essentially equal to the wave height (H), or, let us arbitrarily say in this example, 30 cm (a little less than 1 foot). Calculate the orbital diameter at the depths indicated in Table 11.1.

It is sometimes stated that orbital motion dies out at water depths deeper than one-half the wavelength. Do you still think that this is rigorously true?

____

Explain:

____

____

____

____

____

____

____

____

If the water depth is great enough so that the orbital motion at the bottom of the lake or ocean is very small, the wave passes over the bottom without a great deal of interaction with it. The wave

Table 11.1. (Use for Problem 2.)

Depth (cm)	*Orbital diameter (cm)*
0	30
40	
100	
180	

does not "feel" bottom. Such a wave is called a ***deep water wave***. Following the results of Problem 2, waves traveling in water that is deeper than half a wavelength are commonly considered to be deep water waves. When the bottom is sufficiently shallow so that there is significant interaction between the bottom and the orbital motion of the water, the orbital motion can become greatly distorted. This interaction affects many of the features of wave motion, including the speed at which it travels. Waves that are greatly affected by interaction with the bottom are called ***shallow water waves***. The definition is arbitrary, but waves in water shallower than 1/20 wavelength are commonly considered to be shallow water waves. Waves that exhibit relatively small amounts of interaction with the ocean bottom, and hence have their features only modestly affected, are sometimes called ***transitional waves***. It is clear that whether a wave is a deep water wave or a shallow water wave depends both on the water depth and on the wavelength of the wave. Tides (which travel across the oceans to some extent as waves) and tsunamis (seismic sea waves) may have wavelengths of tens or hundreds of kilometers or even more, and therefore interact with the bottom of even the deepest parts of the ocean. They therefore can be considered to be shallow water waves everywhere in the oceans.

Wave Height: Waves are generated when the wind blows over open water. Three factors affect the height of the waves that are generated. Waves become higher 1) the faster the wind is blowing, 2) the greater the wind duration (i.e., the longer the time that the wind has been blowing), and 3) the greater the distance of open water over which the wind has been blowing (called the ***fetch***). But wave heights will not increase without limit as the wind duration or the fetch increases. For any given wind velocity there is a maximum height to which waves can grow. A longer fetch or a longer time during which the wind blows will result in the formation of more waves, but not higher waves. We describe the condition in which the wave height is the maximum for a given wind velocity as a ***fully developed sea.***

Figure 11.3 is a graph that can be used to estimate the heights of waves generated by winds of a given speed (in this case 54 km/h or about 34 mph) blowing for different lengths of time and over different fetches. A similar graph could be generated for any wind velocity desired.

Figure 11.3 is a composite of many curves, one for each wind duration indicated (i.e., 5 h, 10 h, 15 h, etc.). When the line for a given wind duration becomes horizontal it indicates that an increase in fetch does not result in increased wave height for the wind duration indicated.

The graph shows, for example, that if the 54 km/h wind blows for 5 hours (or longer) over relatively short fetches (less than approximately 50 km or 30 miles) the wave height increases as the fetch increases, reaching a maximum (for 5 hours) of about 2.6 meters when the fetch becomes as great as 50 km. Waves will grow to no greater average height during a period of 5 hours, no matter how long the fetch.

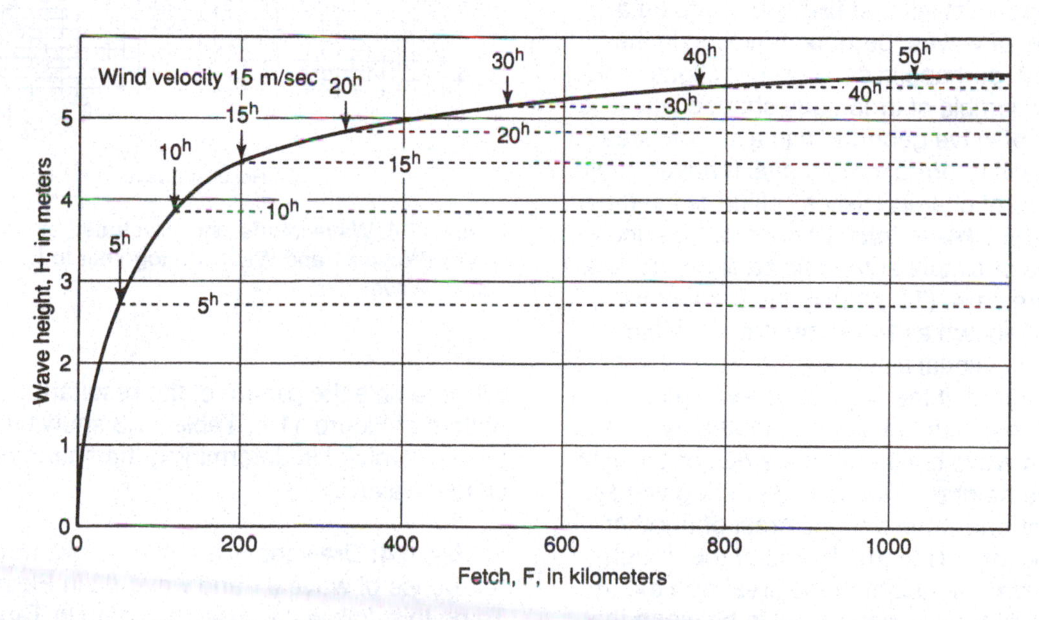

Figure 11.3. Nomogram showing the relationship between fetch, wind duration, and wave height for a wind velocity of 15 m/sec (54 km/h or 33 miles per hour). A comparable graph could be drawn for any other wind velocity. Note that by 50 hours duration a wind of 54 km/h blowing over a fetch of 1000 km or more results in a *fully developed sea*. That is, a greater duration and a greater fetch does not result in increased wave height. (Adapted from H.U. Sverdrup, M.W. Johnson, and R.H. Fleming, *The Oceans*, fig. 134, © Prentice-Hall, Inc., 1942.)

Table 11.2. (Use for Problem 3.)

Length of time (hr)	*Wave height (m)*
5	
10	
15	
20	
25	

In another example we can see that when the fetch is long, (longer than about 1000 km) the wave height is significantly greater when the wind blows for 10 hours than when it blows for 5 hours, but that increasing the time the wind blows from 40 hours to 50 hours makes only a very small change in the wave height.

Problem 3: If the wind blows over a fetch of 200 km with a speed of 54 km/h, for each of the lengths of time indicated in Table 11.2, use Figure 11.3 to determine the wave heights that will develop. List your answers in Table 11.2.

Wave Interference: Frequently waves on the surface of the open ocean do not resemble a single, simple set of waves traveling in a single direction. They may appear (at least at first glance) to be a chaotic jumble of waves, perhaps traveling in the same direction, perhaps in different directions. This sort of chaotic jumble of irregularly shaped waves is characteristic of wave generation in a stormy area. Such a jumble is referred to as a *sea*. It arises when waves of different characteristics collide or interfere with one another. ***Wave interference*** can be understood in terms of simple arithmetic addition, as illustrated in Figure 11.4. (To simplify the presentation, the waves are shown as two-dimensional rather than three-dimensional features.)

Simply stated, if the crest of one wave coincides (or collides) with the crest of another wave the result will be a wave crest that has a height equal to the sum of the heights of the two interfering waves. If the trough of one wave and the crest of another wave coincide (or collide) the height of the trough will subtract from the height of the crest and the resultant wave will have a water level in between that of the colliding crest and trough. For example, on the graph in Figure 11.4 are drawn snapshots of two waves taken at the same time. When the two waves interfere with one another, the water surface

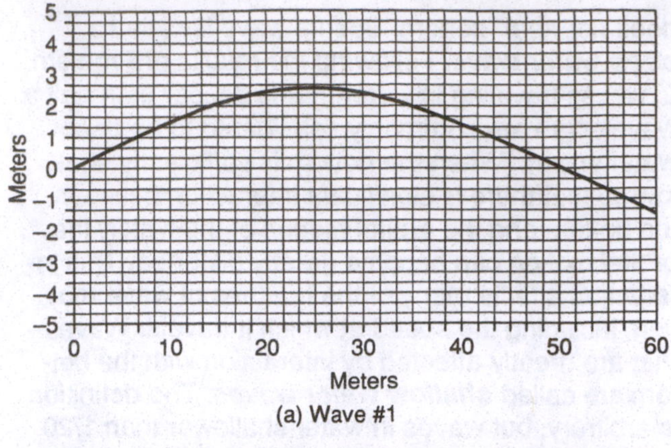

(a) Wave #1

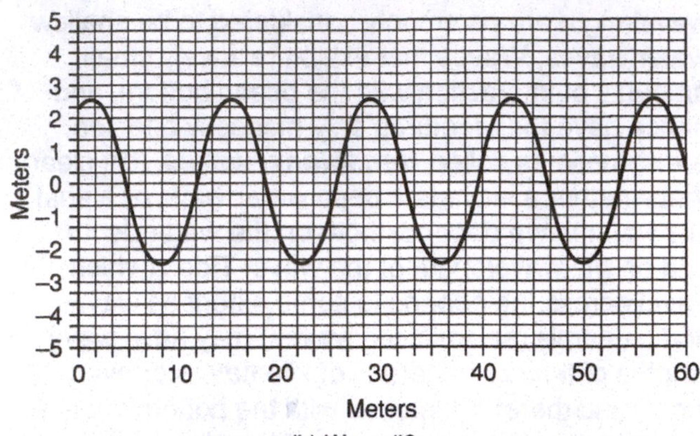

(b) Wave #2

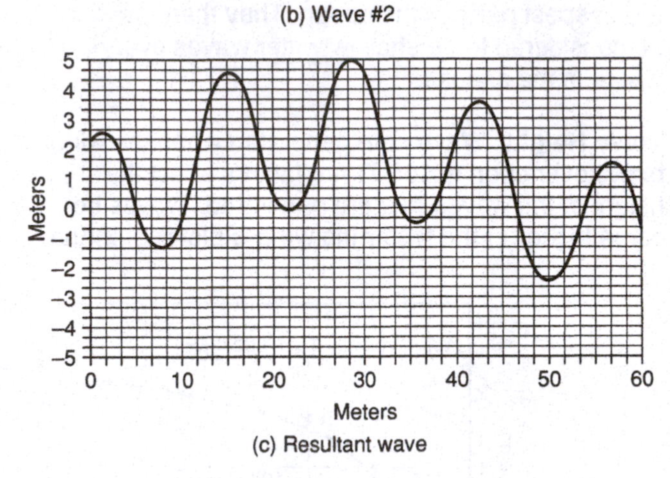

(c) Resultant wave

Figure 11.4. Wave interference illustrated by adding two waves (Wave #1 and Wave #2) together to form the Resultant Wave.

will resemble the picture of the resultant wave at the bottom of Figure 11.4. Table 11.3 shows the calculations involved in determining the shape of the resultant wave.

Problem 4: Draw the wave that results from the interference of wave #1 and wave #2 in Figure 11.5. To do this, follow the example given in Table 11.3. Enter the data in Table 11.4 and plot the resultant wave in the space provided on Figure 11.5.

Your instructor may ask you to do this problem using the computer instead of doing it graphically by

Table 11.3. Example of the calculation of the shape of a resultant wave formed by the interference of two waves (waves #1 and #2 of Figure 11.4).

Horizontal distance (m)	*Height of wave #1 (m)*	*Height of wave #2 (m)*	*Height of combined wave (m)*
0	0	2.3	2.3
10	1.5	-1.6	-0.1
20	2.4	-1.6	0.8
30	2.4	2.3	4.7

Note: The heights of wave #1 and wave #2 are measured on the graph (Figure 11.4). The height of the combined wave is the sum of the heights of wave #1 and wave #2.

hand. (This can be done readily using a spreadsheet program.) The equation of wave #1 is:

$$\text{Height} = 2.5 \times \sin\left(\frac{2\pi \times \text{distance}}{40}\right)$$

The equation for wave #2 is:

$$\text{Height} = 2.5 \times \sin\left(\frac{2\pi \times \text{distance} - 7}{22}\right)$$

Wave Dispersion: In deep water, the velocity at which a wave travels depends solely upon its wavelength. The velocity can be calculated using the formula

$$\text{Velocity (m/sec)} = \sqrt{\frac{gL}{2\pi}}$$

where g = the acceleration due to gravity[1] (approximately 980 cm/sec/sec) and L = wavelength (in meters)

or

$$\text{Velocity (m/sec)} = 1.25\sqrt{L}$$

(1 m/sec = 3.6 km/hr = 2.23 miles/hr)

[1] The acceleration due to gravity, *g*, is the rate at which a falling object accelerates (picks up speed) as it is pulled downward by the earth's gravitational field. Ignoring air resistance, if we dropped a rock or other object from a high cliff, it would be traveling downward at the rate of 980 cm/sec at the end of the first second, 1760 cm/sec (i.e., 2 x 980) at the end of the second second, etc.

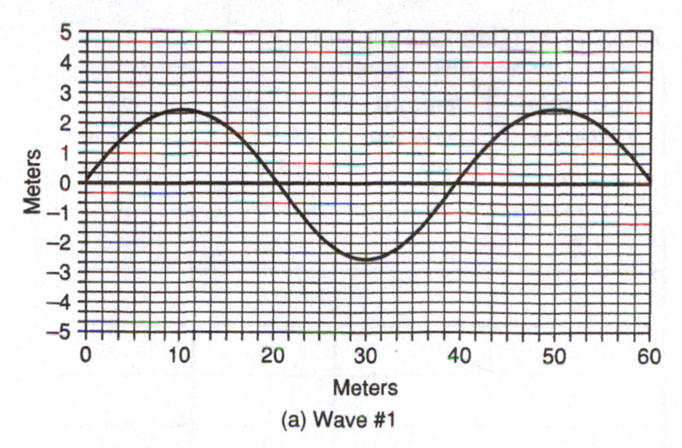

(a) Wave #1

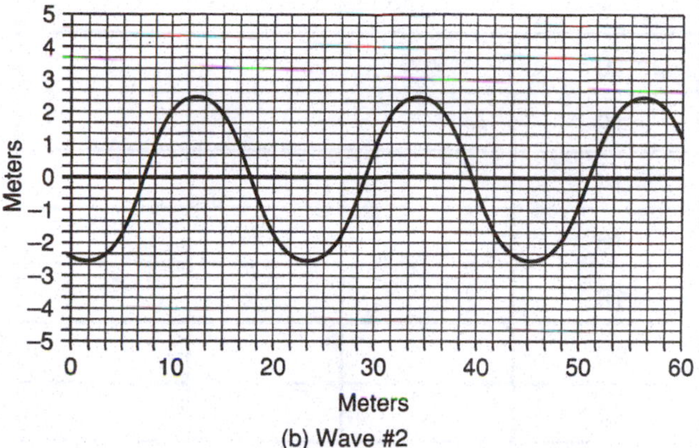

(b) Wave #2

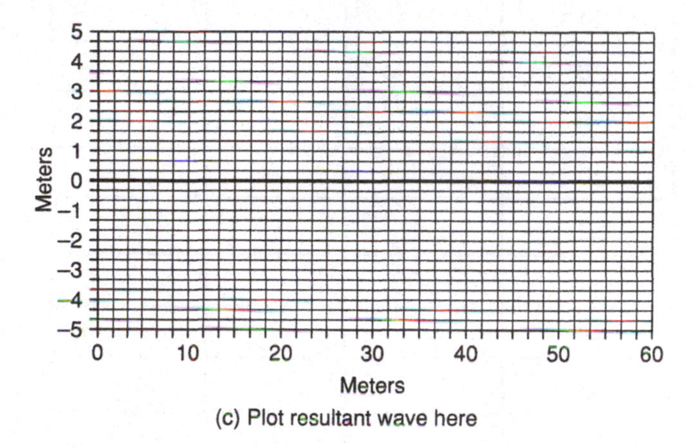

(c) Plot resultant wave here

Figure 11.5. Graphs for Problem 4.

It is useful at this point to consider how waves travel across the water when the surface is disturbed. Imagine that you throw a stone into a still pond. A series of circles spreads out from the spot where the stone hits. If you were to watch very carefully, using a slow motion camera, you would see something that might surprise you. As illustrated in Figure 11.6, individual waves travel faster than the disturbance itself. As individual waves move to the front of the disturbance they disappear. At the same time, new waves develop in the rear of the disturbance.

The speed at which the disturbance moves is called the ***group velocity***. The velocity of individual

Table 11.4. (Use for Problem 4.)

Horizontal distance (m)	*Height of wave #1 (m)*	*Height of wave #2 (m)*	*Height of combined wave (m)*
0			
2			
4			
6			
8			
10			
12			
14			
16			
18			
20			
22			
24			
26			
28			
30			
32			
34			
36			
38			
40			
42			
44			
46			
48			
50			
52			
54			
56			
58			
60			

waves is called the ***wave velocity***. Careful measurements would show that the wave velocity is 2 times the group velocity. The velocity of the group is half the velocity of the individual waves that make it up.

Remember that the waves generated in a storm are a jumble of waves of different heights and wavelengths. That is, they are a ***sea***. Waves with longer wavelengths have higher velocities than waves with shorter wavelengths. The waves of different wavelengths therefore travel away from the storm center at different velocities. As a result, as the waves travel away from the storm center where they were generated, the chaotic jumble of the sea begins gradually to sort itself out. After a few days, the groups of waves with longer wavelengths will have moved significantly farther from the storm center than will the groups of waves with shorter wavelengths. The effects of wave interference will begin to be undone. This phenomenon is called ***wave dispersion***.

Imagine that you are standing on a beach watching the waves come in to shore. On a particular day you see that the waves are long wavelength waves with smooth profiles and gentle slopes. Such waves are called ***swells***. Because you understand wave dispersion you realize that these waves must have traveled a very long distance from the storm center where they were generated, and in doing so they must have left the waves with shorter wavelengths behind. Twelve hours later you return to the same spot on the beach and observe as similar scene, but the wavelength of the swell is somewhat shorter than it was in your earlier observation. These waves with shorter wavelengths originated in the same storm as those with longer wavelengths, but they took longer to reach the spot where you are standing because they traveled more slowly. Knowing the way that wave (and group) velocity varies with wavelength, you can calculate the distance the waves traveled from the storm center where they were generated. (It is said that ancient Polynesians were adept at interpreting the swell in this way, and thereby heeded the warning of impending storms that had not yet struck their islands.)

Problem 5: You visit the beach on a tropical Pacific island at 6 A.M. and find that gentle swells are arriving at the shore with a wavelength of 60 meters. Using the equations on the previous page, calculate the velocity of those waves.

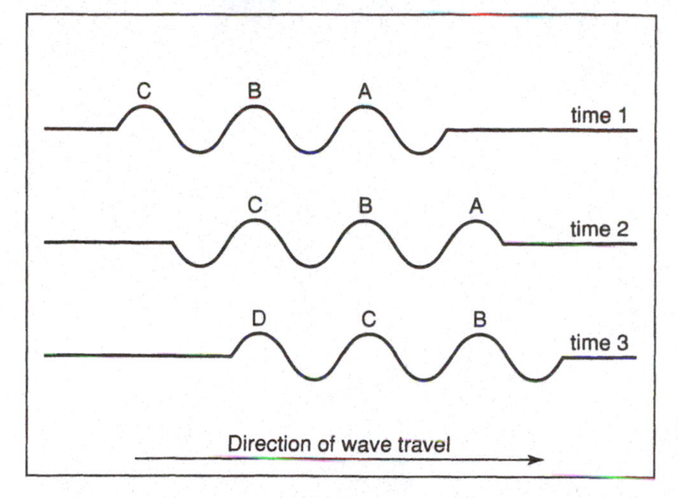

Figure 11.6. A series of cross-sections of a group of three waves traveling from left to right. Individual waves are labeled with letters. Note that at time 2 (one wave period after time 1) wave crest A has moved to the right one full wavelength, but the disturbance has moved ahead only half a wavelength. The trough to the right of wave crest A has disappeared and a new trough has formed at the rear of the group, to the left of wave crest C. At time 3 (one wave period after time 2) the disturbance has moved another half wavelength to the right while the individual waves have moved another full wavelength. The crest of wave A has completely disappeared and a new crest, wave D, has formed at the rear of the group. The velocity of the group is one-half the velocity of individual waves.

What is the group velocity of the group of waves of that wavelength?

You return to the same spot at 6 P.M. on the same day and find that gentle swells are still arriving, but the wavelength is now 40 meters. Calculate the velocity of those waves in km/h.

What is the group velocity of the group of waves of that wavelength?

Remember that distance traveled equals velocity x travel time. In the case of the waves traveling from the storm center to the observation point, we are interested in the group velocity, or the velocity of the disturbance.

$$\text{Distance} = \text{group velocity} \times \text{time}$$

Problem 6: If both groups of swells originated at the same storm center, **calculate how far away that storm center is**. Hint: Assume that both sets of swells left the same point of origin at the same time, so that each group traveled the same distance but at a different velocity.

$$\text{Distance}_{slow} = \text{Distance}_{fast}$$

where Distance_{slow} is the distance the slow (shorter wavelength) wave has traveled from the storm center and Distance_{fast} is the distance the fast (longer wavelength) wave has traveled.

You also know that $\text{time}_{slow} = \text{time}_{fast} + 12$ hours (the length of time between 6 P.M. and 6 A.M. on the day you observed the wave). Therefore,

$$\text{Group velocity}_{fast} \times \text{time}_{fast} = \text{Group velocity}_{slow} \times (\text{time}_{fast} + 12)$$

Solving this equation for time_{fast} we get

$$\text{time}_{fast} = \frac{12 \times \text{Velocity}_{slow}}{\text{Velocity}_{fast} - \text{Velocity}_{slow}}$$

where the velocities are the group velocities you calculated above. Knowing the time and the velocity, calculate the distance to the storm center.

Name: ______________________________ Date: ______________ Section: ____________

Exercise 12
WAVE TANK EXPERIMENTS

Purpose: In this experiment you will gain experience making measurements of waves that are generated in a laboratory apparatus called a ***wave tank***. You will use those measurements to increase your understanding of how waves behave. (For classes that do not have access to a wave tank, photographs are included with this exercise which will permit you to make the necessary measurements.)

The wave tank: A wave tank (Figure 12.1) is a long tank equipped with a paddle that oscillates back and forth at one end. The paddle creates waves that travel the length of the tank. The wave tank is convenient for observing the behavior of simple waves in the laboratory. Using it, we can demonstrate features of waves that are difficult to visualize from observations of the complex wave sets frequently seen on the ocean or on large lakes.

In doing this experiment, it is most convenient to make Polaroid™ photographs of the waves in the tank, and to make measurements directly on the photographs. The wave tank illustrated in the photographs in this exercise is equipped with a horizontal scale, marked at intervals of 10 cm (0.1 m). The wave tank your class uses may have a different scale marked upon it.

Simple wave properties: Every stroke of the paddle in the wave tank creates a new wave. A ***wave period*** is the length of time between the passage of one wave and the passage of the equivalent part of the next wave (e.g., the time between the passage of wave crests). The rate at which waves pass any point is the ***wave frequency***. The frequency, *f*, with which waves pass a point, is the reciprocal of the period, *P*. In other words

$$f = 1/P$$

If a wave has a period of 8 seconds, its frequency is 1/(8 sec) or 0.125 sec^{-1} (sometimes read 0.125 per second).

The ***wavelength*** of a wave, *L*, is the distance from one point on a wave to the equivalent point on the next wave. This means that the wave period is the length of time it takes the wave to travel the distance *L*.

The wave velocity is the speed at which any part of the wave travels. For any wave, the formula below must hold.

$$\text{Velocity} = L/P$$

As seen in Exercise 11, the velocity of a ***deep water wave*** (a wave that does not interact significantly with the bottom) can also be calculated from the wavelength and the physical principles that govern wave travel. For such a wave,

$$\text{Velocity} = 1.25\sqrt{L}$$

Problem 1—Determination of wavelengths and wave velocities: Your instructor will show you how to adjust the speed of the paddle, and will tell you what paddle speed to choose to generate your first set of waves. (If you do not have access to a wave tank, answer the problem using the photographs in

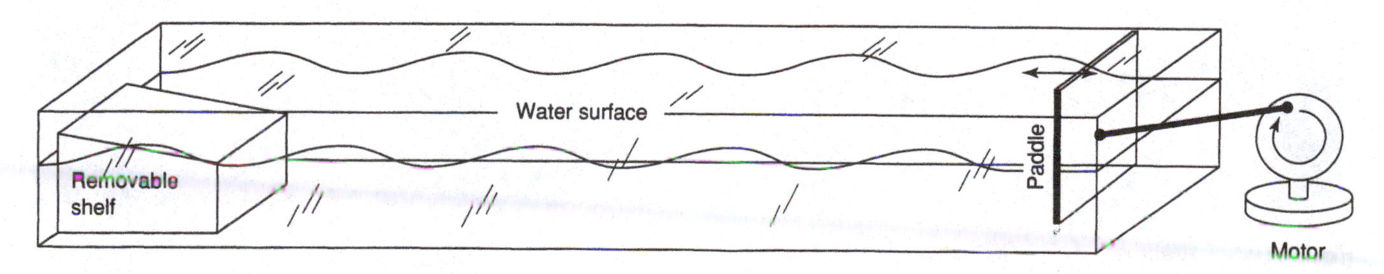

Figure 12.1. A wave tank. The paddle on the right oscillates back and forth, setting in motion waves that travel across the tank from right to left. The speed at which the paddle oscillates is adjustable. The sloping shelf is removable, and when in place it permits simulation of waves approaching the shore.

Figure 12.2. Experiment 1. P = 0.97 sec.

Figure 12.3 Experiment 2. P = 0.85 sec.

Figure 12.4. Experiment 3. P = 0.76 sec.

Figure 12.5. Experiment 4. P = 0.72 sec.

Figures 12.2 through 12.5 and the following information:

Experiment No.	*Period (sec)*
1	0.97
2	0.85
3	0.76
4	0.72

As you proceed through this exercise, enter your answers in Table 12.1. For the first set of measurements, do the following:

1. Set the paddle speed as instructed. Measure the wave period by counting the number of paddle oscillations in 60 seconds.

$$\text{Period} = \frac{\text{60 seconds}}{\text{No. of oscillations}}$$

   Enter the answer in row 1 of Table 12.1. (If you are using the photographs provided, enter the period listed above for Experiment No. 1.)
2. Calculate the wave frequency and enter the value in the table.
3. Take a Polaroid™ photograph of the tank. Using the scale affixed to the tank, measure and record the wavelength of the first wave you can measure at the paddle end of the tank. It is usually easier to measure the distance from wave crest to crest, but in some cases it may be necessary to measure from trough to trough. (In Figures 12.2 through 12.5, vertical lines have been drawn on the photographs to mark convenient points between which you can measure. **The scale divisions marked along the top of the tank in the photographs are 0.1 m apart.**) Record all measurements in Table 12.1.
4. Calculate the velocity of the wave at the paddle end using the general formula, Velocity = *L/P*. Then calculate the expected velocity for a deep water wave of the same wavelength (Velocity = $1.25\sqrt{L}$). Record both values in the table. Compare the velocities you calculated using the two approaches. If they vary by more than 0.2 meters/sec, you should recheck your measurements and your arithmetic.
5. Now measure the wavelength of the last wave measurable at the end of the tank where the waves break. (Deciding where to locate a crest or a trough at this end of the tank can be difficult. Do the best you can.) Record the wavelength.
6. Calculate the velocity of this wave using the general formula, Velocity = *L/P*. Record that value. (Note that you are not asked to calculate the velocity of the wave at the breaking end of the tank using the formula Velocity = $1.25\sqrt{L}$. That formula is applicable only for deep water waves. It is not applicable when the water is shallow compared to the wavelength, as discussed in Exercise 11.)

Repeat steps 1 through 6 for three other wave periods. Follow the directions of your instructor as to the rate of oscillation of the paddle or use the data and photographs provided in this exercise.

Problem 2: Did you note any difference between the *wavelength* at the paddle end of the wave tank and the *wavelength* at the breaker end for any of the waves? If so, why? Did you observe the great-

Table 12.1. Write answers to Problem 1 here.

			Paddle End of Tank			*Breaker End*	
Experiment	*Period [P] (sec)*	*Frequency* (sec^{-1})	*Wavelength [L] (m)*	*Velocity L/P (m/sec)*	*Velocity* $V = 1.25\sqrt{L}$ *(m/sec)*	*Wavelength [L] (m)*	*Velocity L/P (m/sec)*
1							
2							
3							
4							

est differences for the waves with shorter wavelengths or the waves with longer wavelengths? Why?

Problem 3: What, if any, differences in *wave velocity* did you observe between waves at the paddle end of the tank and waves at the breaker end of the tank?

Problem 4: Describe any differences you observed between the *shapes of waves* at the paddle end of the tank and the shapes of waves at the breaker end of the tank.

Standing waves: You have certainly observed the phenomenon of water sloshing back and forth in a swimming pool or in a tray you are carrying. This sloshing is a wave phenomenon which differs from the waves we have observed in the wave tank because of the way in which the waves reflect off the sides of the container. ***Standing waves*** develop when wave systems with identical characteristics encounter one another while traveling in opposite directions. They are called standing waves because the crests and troughs do not move. They remain stationary. Examples of standing waves are shown in Figure 12.6.

Standing waves with one node are found in restricted embayments and in some lakes. Such a wave is called a seiche. The period of a seiche in a rectangular tank is

$$P = \frac{2 \times \text{basin length}}{\sqrt{g \times \text{basin depth}}}$$

where all measurements are in meters, and g is the acceleration due to gravity (9.8 meters/second^2).

Problem 5: This problem is designed for students with access to a wave tank. Calculate the period of a single node standing wave in your wave tank.

Now adjust the paddle movement in your wave tank so that you form the best one-node standing wave you can. Measure the period of that wave. How does it compare to the calculated period?

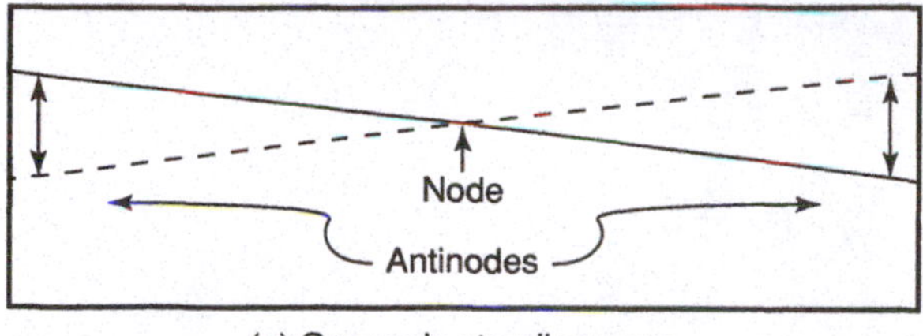

(a) One-node standing wave

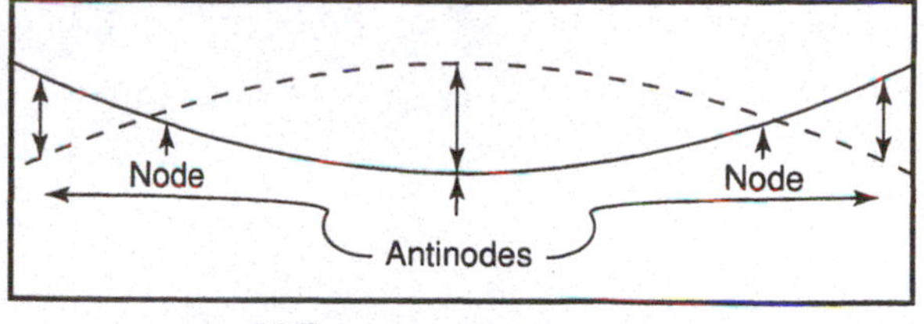

(b) Two-node standing wave

Figure 12.6. Standing waves with one node and two nodes in a tank. In each case, the solid line shows the shape of the water surface at one time. The dashed curve shows the surface one half period later.

____

____

____

____

____

____

____

You can create standing waves in your wave tank with more than one node. The theoretical periods of these standing waves can be readily calculated. The calculations predict that a standing wave with two nodes in a rectangular tank should have a period half that of a wave with a single node, a standing wave with three nodes should have a period one third of a wave with a single node, etc.

Expressed mathematically,

$$P = \frac{2 \times \text{basin length}}{n \times \sqrt{g \times \text{basin depth}}}$$

where n is the number of nodes.

By adjusting the paddle speed, determine the periods, in your wave tank, of standing waves with 2 nodes, 3 nodes, and 4 nodes. List the periods in Table 12.2.

Differences between the calculated and measured periods of standing waves are the result of friction between the water and the walls of the tank.

Table 12.2. For use in Problem 5.

Number of Nodes	*Calculated Period (seconds)*	*Measured Period (seconds)*	*Ratio Measured/ Calculated*
1			
2			
3			
4			

Name: ______________________ Date: ______________ Section: ____________

Exercise 13
COASTAL OCEANOGRAPHY AND SHORELINE EROSION

Purpose: In this exercise you will become familiar with a number of the factors that shape the shores of oceans and large lakes and that affect the people and the property along the shore.

Introduction: Coastal zones are dynamic systems. For example, even when a beach appears much the same from day to day, careful observations and measurements usually indicate that it is undergoing constant change. While the size and shape of the beach may not change greatly, the materials that make up the beach are normally in motion. They are moved along primarily as the result of the waves that hit the beach and the currents produced by those waves. In one sense, a beach is analogous to a river. While a river may appear the same from day to day, the water we see in the river today is not the same water we looked at yesterday. Similarly, while a beach may appear the same this summer as it did last summer, the sand we walk on today may be very different from the sand we walked on last year. Just as water in a river flows downstream, sand moves downdrift along the shore.

Nomenclature of the coastal zone: The ***coastal zone*** is the region, offshore and onshore, in which the waves, tides, and longshore currents affect the rocks and sediments. It is convenient to divide the coastal region into a series of zones on the basis of the processes that are dominant in shaping each zone. This is shown in Figure 13.1.

Beach materials: Beaches are made of whatever loose material is available at the coastline. Most of us think of beaches as made up of ***sand***. In fact, while sand is the most common beach material, beaches can be made up of sedimentary particles of almost any size, from ***cobbles*** to ***pebbles*** to sand to silt and even ***clay***. (See Exercise 6 for a review of these terms.)

The processes of sediment transportation play a major role in determining the particle size distribution of the materials that make up the beach. Generally, the more rapidly water moves, the coarser is the sediment that it is able to carry and the coarser is the sediment that can be eroded. This is illustrated in Figure 13.2.

Sediment is moved along shores primarily by water, both by currents in the breaker zone and by the swash and backwash of waves along the beach face. (Wind, of course, can move sediment too, as anyone who has picnicked on the beach on a windy day can testify, but movement by water is usually more important than movement by wind.)

The material on a beach becomes ***sorted*** by the movement of waves back and forth on the beach and the movement of currents along the beach. At any spot on a beach you will frequently find that the material is ***well sorted***. That is, most of the material falls within a fairly narrow size range, i.e., fine sand, coarse sand, etc. This is because the water is not moving fast enough to transport coarser material than that of the dominant size range to the

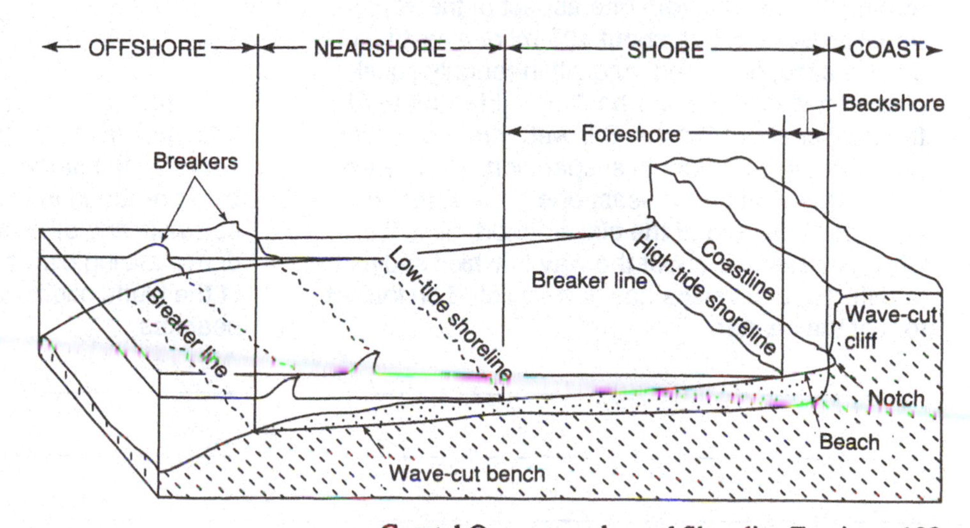

Figure 13.1. Terminology of the coastal zone. The ***offshore*** zone is the zone beyond the breaking waves. The ***nearshore*** extends from the low tide line outward to the zone of breaking waves. The ***shore*** consists of the ***foreshore*** (which is exposed at low tide and submerged at high tide) and the ***backshore*** region (which extends from the high tide line to the coastline). The ***coastline*** is the most landward extent of direct erosion by ocean waves. Two ocean surfaces are shown, one high tide and one low tide.

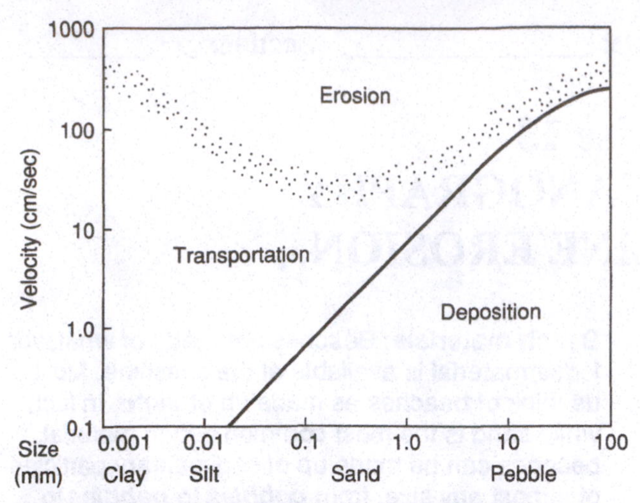

Figure 13.2. Results of an experimental study by Hjulstrom that shows that: 1) for any given particle size, the water velocity necessary to erode the material is greater than that necessary to transport it; 2) if material is being transported and the water velocity slows sufficiently the material will be deposited; and 3) for sand and coarser particles, the range of water velocities capable of creating a beach of a particular particle size is fairly small. Look, for example, at sand with a particle size of 1 mm. If the current is traveling more slowly than about 9 mm/sec, particles of this size will settle out before they get to the beach site. They will not reach the beach. If the current is traveling more rapidly than about 50 cm/sec, a beach made up of particles of this size will be eroded and the sand will be carried away. (Reprinted with the permission of Macmillan Publishing Company from *The Earth,* Fourth Edition, (fig. 10.15), by Edward J. Tarbuck and Frederick K. Lutgens. Copyright © 1993 by Macmillan Publishing Company.)

beach, but it is moving sufficiently rapidly to ***winnow***, or wash away, sediment that is much finer than the dominant grain size.

Problem 1: You can quickly see the particle size sorting that results from one aspect of the depositional process. Place about 100 ml of a mixture of coarse sand, fine sand, and silt in roughly equal proportions in a one liter beaker. Add water to fill the beaker about 3/4 full. Stir well with a spoon to put all of the sediment in suspension. Then leave the beaker alone for at least one hour. After an hour, or at the end of the class period, describe what you can see about the way in which sediment of different grain sizes has accumulated on the bottom of the beaker.

A second factor, in addition to the speed at which water is moving, that helps determine the sizes of grains that make up a beach, is the nature of the source(s) of the material. In many parts of the United States, particularly along much of the Pacific coast and much of the Gulf of Mexico, much of the sediment that forms the beach has been carried to the coast by rivers. Along much of the Atlantic coast, on the other hand, beaches consist of materials that accumulated offshore during lowstands of sea level that occurred each time the large icecaps covered the continents. An additional source, especially in the southeastern United States, is ground-up shells and other biological debris. Along many parts of the Great Lakes the primary source of beach material is the eroding ***bluffs*** (low cliffs) at the water's edge.

Problem 2: In many parts of the coast of southern California beaches have become smaller in recent years. This has resulted, at least part, from the damming of mountain streams for purposes of flood control. Explain why this might cause the beaches to become smaller.

Along the south shore of Lake Erie the beaches have also become smaller in the past 40 years. In the same interval large sections of the bluff bordering the lake have been protected by seawalls and other structures built for the purpose of minimizing bluff erosion. Explain how protection of the bluffs might result in shrinkage of the beaches.

Longshore currents and longshore drift: The interaction of waves with the shore typically results in a ***longshore current*** flowing parallel to the coastline in the breaker zone. This happens whenever the waves strike the shoreline at an oblique angle (i.e., when the waves are not parallel to the shoreline). The establishment of a longshore current is illustrated in Figure 13.3. Each time a wave breaks, its orbital motion (see Exercise 11) is converted to translational motion. The water moves in the direction in which the wave is traveling. As the wave surges toward the beach the water moves in the direction that the wave moves. This is approximately perpendicular to the shoreline, but with some motion in the longshore direction as well. As the surf rushes back out, the water maintains some of its momentum in the longshore direction. With each breaking wave, the water moves a little farther along the shore. The transport of sand along the shore in the direction of the longshore constitutes part of the ***longshore drift***.

In a somewhat similar fashion, sand is transported along the beach above the shoreline. With every incoming wave, water rushes up the beach as ***swash***. Some of the swash percolates down through the sand and leaks back to the ocean. And some of the water rolls back down the beach to the ocean as ***backwash***. The swash and backwash follow a zigzag pattern similar to the pattern shown in Figure 13.3 for the nearshore breaker zone. It is very easy to see this part of the longshore drift if you stand at the water's edge and look at a single

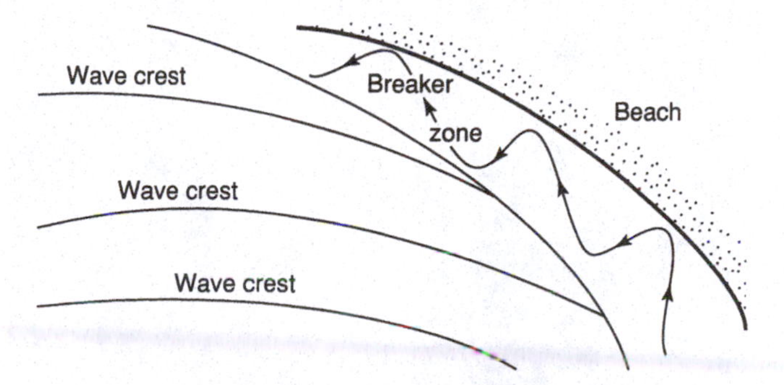

Figure 13.3. A longshore current develops in the surf zone when waves approach the shoreline at an oblique angle. The longshore current is shown by the zigzag curve with the arrows.

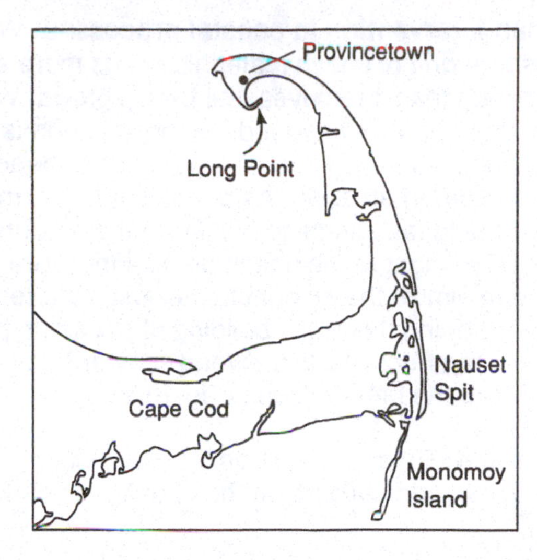

Figure 13.4. Map of Cape Cod, for use in Problem 3.

distinctive shell or sand grain in the swash zone.

The direction of longshore drift depends upon the direction from which waves are arriving at the beach. This may vary from day to day, although along most stretches of shoreline there is a predominant direction of longshore drift. This commonly results in the formation of characteristic beach features from which the characteristic direction can be identified.

Problem 3: Figure 13.4 is a map of Cape Cod. Draw the direction of the predominant longshore drift at Long Point, near Provincetown. Draw the direction of predominant longshore drift in the vicinity of Nauset Spit and Monomoy Island. Draw a few wave crests to show the predominant direction of waves that give rise to the drifts that you have drawn in the two regions.

In the space below, indicate why you drew the arrows the way you did.

Human intervention in coastal processes: When longshore drift of beach materials brings more sand to a beach than it removes, the beach grows. When more is lost than is imported, the beach shrinks. Along most stretches of coast, wide stable beaches are considered an asset. As a result, it is common for people to take steps to maximize the amount of material brought to the beach by the longshore drift, and to minimize the amount that is lost. One technique for doing this is the building of ***groins***, typically stone structures that extend seaward from the beach into the breaker zone (Figure 13.5).

Problem 4: Refer to Hjulstrom's curve (Figure 13.2) Why does sand build up on the updrift side of the groin?

Why does the beach typically erode on the downstream side of the groin?

The sand is pulled from that side and theres nothing to stop it

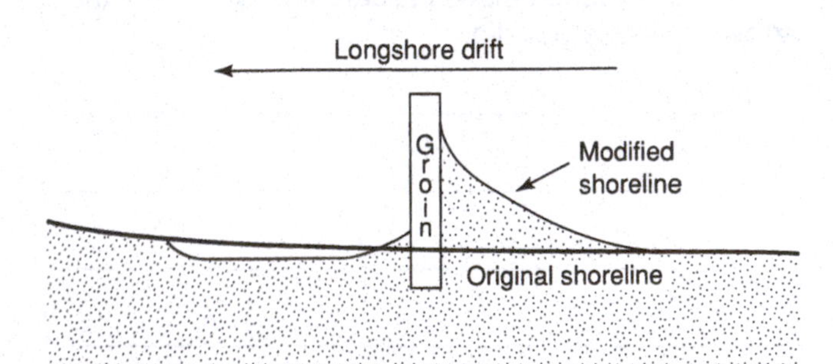

Figure 13.5. The effect of constructing a groin on the shape of a beach. The heavy line shows the shape of the beach before construction of the groin. The lighter line shows the shoreline after construction of the groin. New beach material is deposited on the updrift side of the groin. A small amount of new material sometimes also accumulates immediately downdrift, but erosion of material typically is the predominant process on the downdrift side of the groin.

Why do you think that groins are typically built in groups (see Figure 13.6) rather than singly?

It minimizes the erosion to the max, and so theres a less drastic downstream.

Problem 5: ***Jetties*** are structures similar to groins, but usually much larger. They are commonly built where a river enters the ocean (or large lake) to slow the longshore drift of sediment and thereby prevent the river mouth from becoming choked with sand or silt. It is common for shore erosion to accelerate downdrift from a jetty, as seen in Figure 13.7. Why do you think that this is so?

Figure 13.6. A groin field, or series of groins constructed at Miami Beach. (Courtesy of U.S. Army Corps of Engineers.)

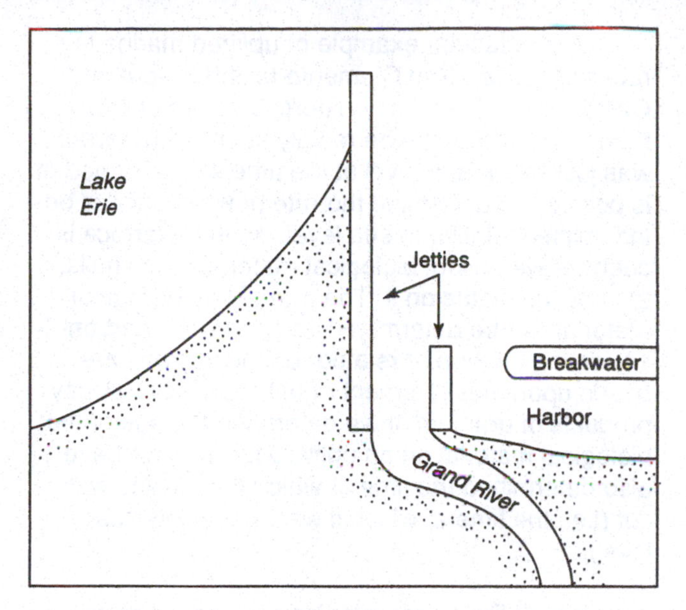

Figure 13.7. Jetties built where the Grand River flows into Lake Erie.

Sea level and shore characteristics: Along much of the coast of the United States, shorelines are retreating (moving landward) at a measurable rate. This is not true everywhere, however, and in some places the shoreline is actively advancing in the seaward direction. Whether the shoreline advances or retreats depends, in part, upon whether the level of the ocean goes up or down relative to the land (Figure 13.8). Because both the level of the land and the level of the sea can change, however, it is not always simple to predict what will happen in the future to the position of the shoreline.

The level of the ocean can change for a number of reasons. Perhaps the most obvious of these is advance and retreat of continental ice sheets. The water that forms these ice sheets has evaporated from the oceans (and falls on the land as precipitation), so as the ice sheets grow, sea level drops. When the ice sheets retreat, as they did most recently beginning at about 18,000 years ago,

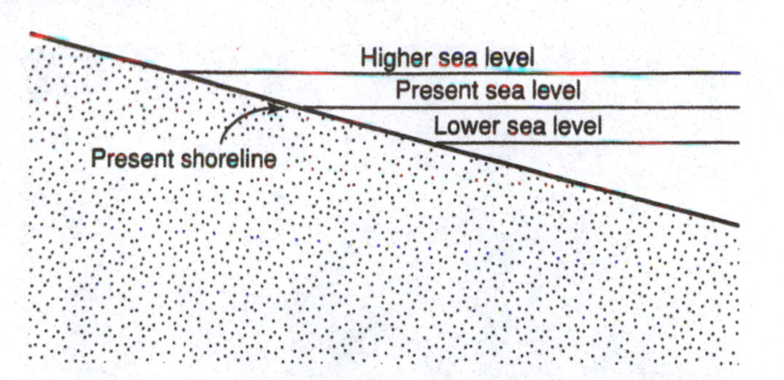

Figure 13.8. Cross section illustrating relationship between sea level and position of the shoreline. Geometry dictates that if the land surface slopes seaward the shoreline must move landward as sea level rises and shoreward as sea level drops.

the water that had been stored on the continents as ice returns to the oceans. Sea level rises. Sea level is now about 120 m (400 ft) higher than it was when the ice sheets were at their maximum extent. Sea level is also affected by the temperature of the water. As the ocean warms, water expands (Exercise 8) and sea level rises. Sea level appears to have risen by about 3 inches over the past 100 years (although this is extremely difficult to measure and there is some uncertainty about the magnitude of the elevation). Many oceanographers believe that this is the result of global warming during that time and the thermal expansion of sea water that resulted. Coastal regions that have been affected by rising sea levels and the submergence by the ocean of land that was once dry are called submergent coasts. Most of the Atlantic and Gulf Coasts of the United States are submergent.

Problem 6: Figure 13.9 is a satellite view of Chesapeake Bay, and the surrounding lands of Virginia, Maryland and Delaware. The body of water at the right of the photograph is the Atlantic Ocean. This region contains many features that clearly indicate that it is a submergent coast. Describe some of the features you see in the photograph that suggest that this coast was shaped as a consequence of submergence of formerly dry land under a rising ocean.

Figure 13.9. Satellite view of Chesapeake Bay and surrounding Virginia, Maryland, and Delaware. The water on the right-hand edge of the photograph is the Atlantic Ocean. (Courtesy of Earth Satellite Corporation.)

Along some coasts tectonic (mountain building) forces have caused the land to move upward at a greater rate than the level of the ocean. The net result is that the coast is emerging from the ocean. Many of the features of an ***emergent coast*** are shaped by the tectonic processes that have caused the region to be uplifted, although these may be modified by marine processes as time passes. Much of the Pacific Coast of the United States is an emergent coast. The cliffs that overlook much of the ocean in this region have been cut and straightened to some extent by the erosive action of the waves.

Uplifted marine ***terraces*** are a common feature of many emergent coasts. Marine terraces are cut by the action of waves. If a terrace has been cut on a coast, and the coast has subsequently been uplifted, the terrace may be exposed well above today's sea level (Figure 13.10).

A spectacular example of uplifted marine terraces occurs on San Clemente Island in southern California (Figure 13.11). There, a series of more than ten terraces can be readily seen. Each terrace was just below sea level at the time it was formed. It is possible to determine the rate at which land is being uplifted relative to sea level. As each terrace is cut by wave action, biological material (sea shells, etc.) accumulates on it. The age of this biological material can be determined using the ^{14}C (carbon-14) dating method or related techniques that are based upon measurement of the radioactive decay products of uranium. If we determine the age of the biological material on an uplifted terrace, we have also determined the time at which the terrace was cut (i.e., the time at which it was just below sea level).

Problem 7: Biological remains on a series of three uplifted terraces in southern California have been dated. A fourth terrace is actively being cut 2 meters below sea level. The ages and the elevations of the terraces above sea level are given in Table 13.1. Each terrace was formed at a time when the size of the continental ice sheets was similar to that of today. (That is, no significant ice sheets covered North America or Eurasia, and large ice sheets covered most of Greenland and Antarctica.) Use the data in the table to calculate the rate (in millimeters per year) of uplift of this region of the coast. Remember, 1,000 millimeters equals 1 meter.

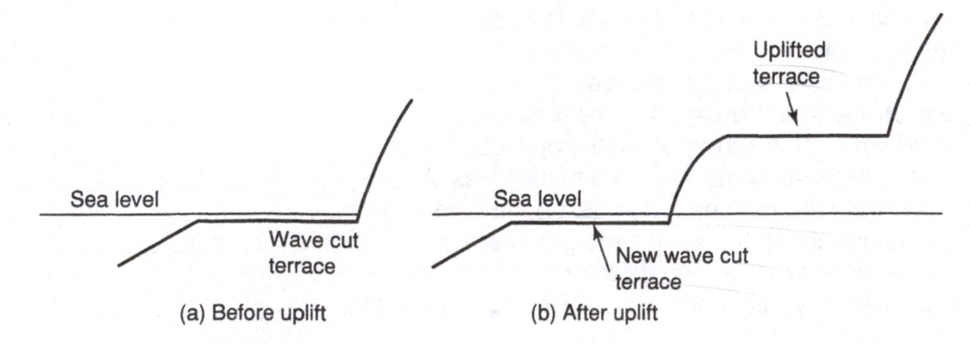

Figure 13.10. Development of an uplifted marine terrace. On the left is a terrace being cut just below sea level by the action of waves. On the right, the land has been subsequently uplifted by tectonic forces. The terrace once below sea level (left) is now exposed above the level of the ocean, while a new terrace is being cut below sea level.

Figure 13.11. Wave-cut terraces on San Clemente Island, California. Each terrace was just below sea level when it was cut. Now the highest terraces are about 400 m (1320 ft) above sea level. (Photograph by John Shelton.)

Table 13.1. For use in Problem 7.

Terrace	*Elevation (m above sea level)*	*Age (years before present)*
1	−2.0	0
2	2.5	8,000
3	45.0	105,000
4	90.0	227,000

Name: ______________________________ Date: _______________ Section: ______________

Exercise 14
THE CARBON CYCLE AND THE GREENHOUSE EFFECT

Purpose: You have undoubtedly read a great deal in the news about the ***greenhouse effect***, carbon dioxide (CO_2), and global warming. The concentration of CO_2 in the atmosphere has increased every year for the past century, and it seems clear that at least much of this increase is ***anthropogenic*** (i.e., caused by human activities). While the effect of this anthropogenic CO_2 on the earth's climate is less well established, it is quite possible that significant warming may result within your lifetime if the present trend of increasing atmospheric CO_2 continues. In this exercise you will develop a greater understanding of some of the scientific issues involved in the current controversy over the present and future role of anthropogenic CO_2 in affecting the world's climate.

The greenhouse effect: Every object gives off, or radiates, a kind of energy we usually describe as ***electromagnetic***. One characteristic of electromagnetic radiation is that it exhibits many properties of waves, properties shared to some extent with ocean waves and seismic waves. We can speak of the wavelength of an electromagnetic wave in the same way we speak of the wavelength of a wave on the ocean (Exercise 11). The wavelengths of radiation given off by an object depend upon its temperature. Most of the radiation given off by the sun is in the ***spectrum*** (i.e., range of wavelengths) we call ***visible light***. Violet is at the short wavelength end of the spectrum and red at the long wavelength end. There is nothing to distinguish visible light from electromagnetic radiation that has shorter or longer wavelengths except that the wavelengths of light are in the range that our eyes are able to see. Electromagnetic radiation with wavelengths shorter than those of visible light is called ***ultraviolet*** and is given off by objects hotter than the surface of the sun. Objects cooler than the sun, like the surface of the earth, give off longer wavelength radiation called ***infrared*** (Figure 14.1). Do not confuse the infrared radiation given off by the earth with sunlight *reflected* from the surface of the earth. Reflected sunlight creates the images we see in photographs of the earth taken from outer space. Our eyes cannot see the infrared radiation given off by the earth, but infrared sensors are able to detect it.

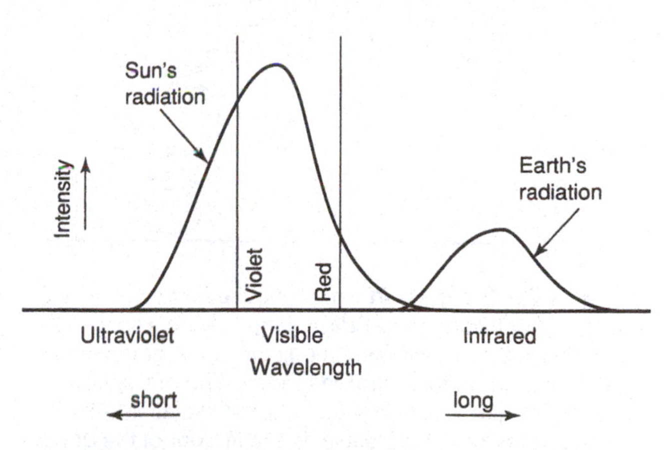

Figure 14.1. The spectrum of electromagnetic radiation given off by the sun consists largely of visible light. Because of its lower temperature, the radiation given off by the earth is in the infrared range.

The earth's atmosphere is, to a large extent, transparent to visible light. That is, a large percentage of the incoming solar energy that is not blocked by clouds passes through the atmosphere and strikes the earth where it warms the surface. The atmosphere is to a large extent opaque to the infrared wavelengths that the earth radiates. That is, much of the outgoing radiation is blocked by the atmosphere, especially by H_2O vapor and CO_2 in the atmosphere. The behavior of the atmosphere has many parallels with that of a horticultural greenhouse (Figure 14.2), and that similarity has given rise to the name ***greenhouse effect***.

The electromagnetic radiation blocked by the atmosphere (or by the glass in the greenhouse) does not disappear. It is converted to another form of energy, heat. Thus warmed, the atmosphere (or the glass) itself radiates electromagnetic energy (Figure 14.3). Eventually, all of the energy that the earth receives from the sun escapes to space. But there is always more energy in the system than there would be if the greenhouse effect were not operative. The earth (or the greenhouse) is warmer as a result. The atmosphere plays a role very much like the blanket on your bed, making it harder for your body heat to escape. Eventually, of course, it does escape.

There are a number of gases in the atmosphere that are important as greenhouse gases

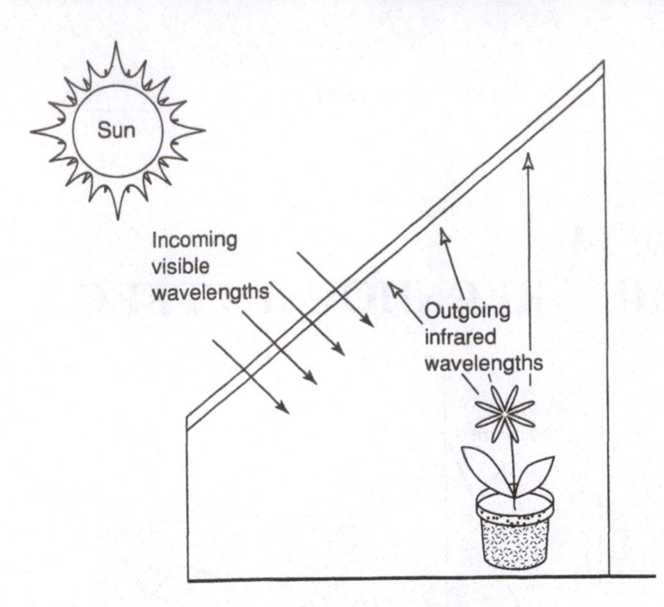

Figure 14.2. The glass in a greenhouse is transparent to much of the incoming radiation (largely visible light). The incoming radiation warms the objects in the greenhouse, and they then radiate infrared energy. The glass, however, is opaque to much of the infrared radiation. As explained in the text, that results in the interior of the greenhouse being much warmer than it would be if this effect were not operative. The earth's atmosphere acts in a fashion similar to the glass in a greenhouse.

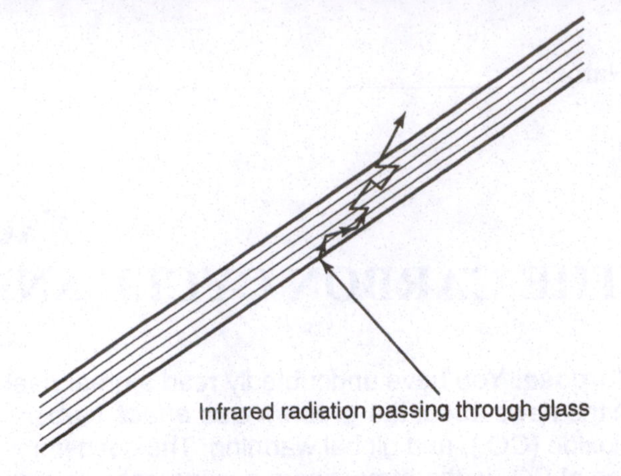

Figure 14.3. As infrared radiation is absorbed by the glass in a greenhouse (or by gases in the atmosphere) it warms the material that absorbs it. That material re-radiates in all directions, and much of the re-radiated energy may be absorbed. Eventually all of the energy escapes, but it remains in the system for a much longer time than it would if the greenhouse effect were not operative.

(i.e., important absorbers of infrared radiation). Most important is H_2O. Also very important is CO_2. Other gases, some anthropogenic, are also significant. Chlorofluorocarbons (Freons), better known for their role in destroying atmospheric ozone, also act as greenhouse gases. Methane (CH_4), which has both natural and anthropogenic sources, is also a greenhouse gas.

There is a great deal of concern that human activities may intensify the greenhouse effect. However, it is important to remember that the greenhouse effect is a natural process. The main greenhouse gas of the atmosphere, H_2O, is abundant in the atmosphere because of the evaporation of water from the oceans and the continents. The second most important greenhouse gas, CO_2, is present in the atmosphere largely as a result of natural processes that we will consider further below. Without a greenhouse effect, the earth would be a cold, probably frozen, planet. The concerns about which we read in the news do not arise because humans have created a greenhouse effect, but because of the fear that they have intensified a natural process.

Carbon dioxide in the atmosphere: CO_2 is an important greenhouse gas in spite of the fact that it is present in the earth's atmosphere in very low concentrations, approximately .03%. In the twentieth century, the combustion of fossil fuels by humans has added significant amounts of CO_2 to the atmosphere. As the result of very careful measurements made for several decades we know that while the concentration of CO_2 in the atmosphere has increased (Figure 14.4), it has not increased as much as one might predict simply from a consideration of the amount of anthropogenic CO_2 produced. In other words, mechanisms in nature operate to remove a part of the anthropogenic CO_2 from the atmosphere.

The carbon cycle: In order to understand the processes that govern the concentration of CO_2 in the atmosphere we will look at the flow of carbon through the atmosphere, biosphere, hydrosphere and lithosphere. This should enable you to gain a feeling for the sensitivity of the atmospheric CO_2 concentration to anthropogenic inputs.

Carbon is one of the primary constituents of all terrestrial life. It enters the biosphere when plants carry out the process of ***photosynthesis***.

$$6CO_2 + 6H_2O \xrightarrow{\text{light}} C_6H_{12}O_6 + 6O_2$$

Plants take CO_2 and H_2O, and using light as an energy source, combine them to make glucose ($C_6H_{12}O_6$, a sugar) and oxygen gas. The CO_2 is one form of ***oxidized*** carbon. The sugar is one form of ***reduced*** carbon. Plants and animals then use the glucose to build other ***organic***, or carbon-containing compounds, all of which can be thought of as reduced. In oxidized carbon compounds the carbon atoms are bonded, at least to a significant extent, to

Figure 14.4. Concentration of atmospheric CO_2 at Mauna Loa Observatory, Hawaii, 1958 to 1988, expressed in parts per million by volume (ppmv).

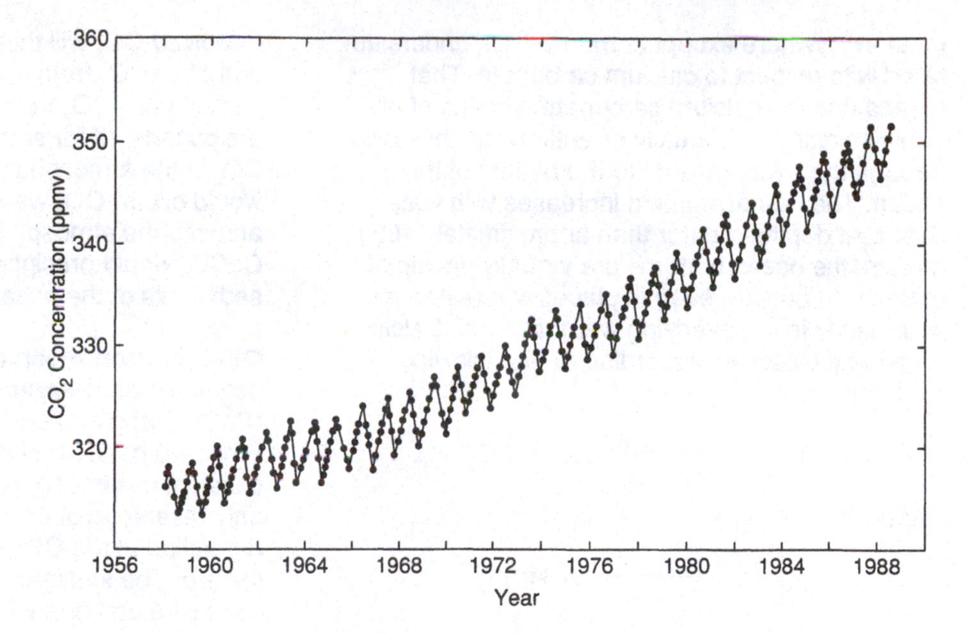

oxygen atoms. In reduced carbon compounds most carbon atoms are bonded to other carbon atoms and/or to hydrogen atoms.

Reduced carbon compounds are not stable in the oxygen-rich atmosphere of the earth, and given sufficient time they will be oxidized back to CO_2, H_2O, etc. In the absence of the activities of bacteria or higher organisms, the oxidation is very slow, however. In contrast, most organisms, bacteria, plants, and animals, are able to carry out rather rapidly the oxidation that would go on only very slowly without their help. Oxidation of organic matter by organisms, called ***respiration***, is in many respects the reverse of photosynthesis. Although almost any organic compound can be oxidized, we will write the respiration reaction for just one, glucose.

$$C_6H_{12}O_6 + 6O_2 \longrightarrow 6CO_2 + 6H_2O$$

The energy released in this reaction is the light energy (normally solar energy) that the plants captured and used to form the chemical bonds of the glucose. It is the energy that we use, for example, to move and to keep warm.

Considered on a global scale, the processes of photosynthesis and respiration go on at almost the same rate. A little more organic matter forms each year by photosynthesis than is destroyed by respiration. That small fraction of the organic matter that is not oxidized becomes buried in sediments. Some of it ultimately becomes coal and petroleum. If that were the end of the story, CO_2 would disappear from the atmosphere and the O_2 concentration in the atmosphere would increase. However, other processes are going on that make the picture much more complex.

Carbon in the oceans: When CO_2 dissolves in seawater it undergoes a series of chemical reactions with the water itself. The first of these is the formation of carbonic acid (H_2CO_3).

$$CO_2 + H_2O \longrightarrow H_2CO_3$$

The carbonic acid then undergoes a series of ***dissociation*** reactions.

$$H_2CO_3 \longrightarrow H^+ + HCO_3^-$$

$$HCO_3^- \longrightarrow H^+ + CO_3^=$$

H^+ or ***hydrogen ion*** is the ion that gives acids their acidic properties. HCO_3^- is called ***bicarbonate*** ion, and $CO_3^=$ is called ***carbonate*** ion. The acidity of the solution determines the extent to which the dissociation reactions proceed. In seawater, carbonic acid dissociates almost completely to bicarbonate ion, but bicarbonate ion dissociates almost not at all to carbonate ion. So in seawater, bicarbonate ion is the dominant ***species*** of oxidized carbon. If the amount of CO_2 in the atmosphere increases, the amount of CO_2 that dissolves in the oceans must also increase. Notice, considering all three of the chemical reactions above, that the effect of dissolving more CO_2 in seawater is not only the production of more bicarbonate ion. The amount of hydrogen ion (or acidity) of the seawater also increases.

In the sediment of the ocean, in addition to organic or reduced carbon there is abundant oxidized carbon. It is present as calcium carbonate (in the form of the minerals ***calcite*** and ***aragonite***), mostly the remains of single cell animals (***foraminifera***) and plants (***coccolithophores***). Seawater is, almost everywhere except at the surface, undersaturated with respect to calcium carbonate. That means that the calcium carbonate remains of organisms dissolve, partially or entirely, as they fall through the water column to the bottom of the ocean. The rate of solution increases with water depth. At depths greater than approximately 4000 meters, the ocean bottoms are virtually devoid of calcium carbonate, even in places where it forms abundantly in the overlying water column. Calcium carbonate dissolves according to the following equations

$$CaCO_3 \longrightarrow Ca^{++} + CO_3^{=}$$

and then

$$CO_3^{=} + H^+ \longrightarrow HCO_3^-$$

The H^+ required to dissolve the $CaCO_3$ forms as a consequence of the dissolution of CO_2 and the subsequent dissociation of the carbonic acid formed, as described above. In turn, the removal of that hydrogen ion by reaction with the carbonate ion produced by calcium carbonate dissolution allows more CO_2 from the atmosphere to dissolve in the oceans.

We may say that the ocean-atmosphere system is **buffered**. That is, mechanisms operate in it that resist, to some extent, changes imposed from outside. In other words, if the amount of CO_2 in the atmosphere increases, some of that CO_2 will dissolve in the oceans. Some of the products of that dissolved CO_2 will then be used up by the dissolution of $CaCO_3$ from the sediment. That, in turn, will permit more CO_2 from the atmosphere to dissolve in the oceans. (If for some reason the concentration of CO_2 in the atmosphere were decreased the reverse would occur. CO_2 would move out of the oceans and into the atmosphere, and, as a result, more $CaCO_3$ would precipitate and accumulate in the sediments of the oceans.)

Other carbon reservoirs: The sediments of the ocean act as ***reservoirs*** of both oxidized carbon ($CaCO_3$) and reduced carbon (organic matter). While we have emphasized the oceans in this discussion, it is important to note that they are not the only reservoirs of carbon that are important to the regulation of the CO_2 concentration in the atmosphere. The sedimentary rocks of the world also constitute an important reservoir. The sedimentary rocks contain both reduced carbon (organic matter including petroleum and coal), and oxidized carbon ($CaCO_3$). When these are weathered at the earth's surface some of the reduced carbon becomes oxidized to CO_2. Prior to the introduction of anthropogenic CO_2, the concentration of CO_2 in the atmosphere was approximately in steady state (Exercise 7). That is, the amount of CO_2 formed by the oxidation of organic matter during weathering of sedimentary rocks was approximately equal to the amount of CO_2 removed from the atmosphere when newly formed organic matter became buried in sediments. It is probable (although not certain) that the CO_2 concentration of the atmosphere has remained

Table 14.1. Data Concerning the Carbon Cycle.

Mass of the atmosphere	5×10^6 Gtons
Concentration of CO_2 in atmosphere at present (expressed as parts carbon per million parts air)	90.0 ppm
Concentration of CO_2 in atmosphere in 1870 (expressed as parts carbon per million parts air)	76.4 ppm
Mass of the oceans	25×10^9 Gtons
Concentration of HCO_3^- in oceans (expressed as units of carbon per million units of seawater)	27.5 ppm
Annual rate of photosynthesis by all plants (as carbon)	60 Gtons/yr
Annual rate of production of anthropogenic CO_2 per year (as carbon)	4.25 Gtons/yr
Total amount of anthropogenic CO_2 added to atmosphere since 1870 (expressed as carbon)	225 Gtons

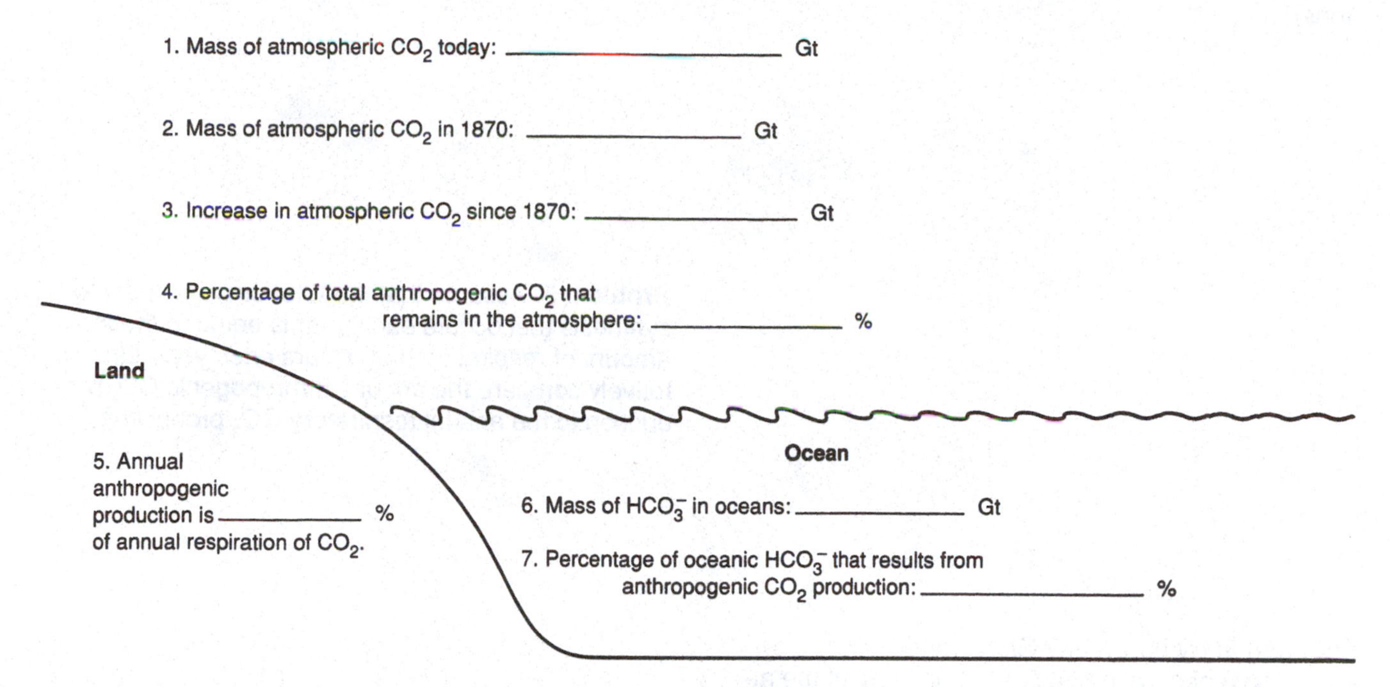

Figure 14.5. Post answers to Problems 1 through 7 here.

within ±50 percent of the modern value over the past several million years.

Another large reservoir of carbon is the ***terrestrial biosphere*** and the remains of recently grown, but not yet decayed, organic matter in soils. To some extent the terrestrial biosphere acts as a buffer. As the CO_2 concentration of the atmosphere increases, the rate of photosynthesis of plants also increases, converting atmospheric CO_2 to reduced carbon at a greater rate.

Quantifying the carbon cycle: Important data pertinent to the carbon cycle of the world and the effect of anthropogenic CO_2 on it are given in Table 14.1. The data will be needed in the problems of this exercise.

Units of ***Gtons*** or ***gigatons*** are 10^{15} grams or 10^9 metric tonnes (approximately 10^9 English tons, or one thousand million tons).

The unit ***ppm*** stands for ***parts per million***. Usually, as in the case of this exercise, ppm are expressed in units of weight. Thus if it is stated that jelly beans in a mixture of candy are present in a concentration of 265 ppm it means that for a million pounds of the candy there are 265 pounds of jelly beans, or equivalently, for a million grams of the candy there are 265 grams of jellybeans. It therefore follows that if we have 821 ounces (oz) of the candy we have

$$821 \text{ oz candy} \times \frac{265 \text{ oz jelly beans}}{1{,}000{,}000 \text{ oz candy}}$$

$$= 0.219 \text{ oz jelly beans}$$

After calculating the answers to problems 1 through 7, below, fill in the appropriate blanks in Figure 14.5.

Problem 1: Calculate the total mass of CO_2 in the atmosphere today (in Gtons of carbon and also in tons).

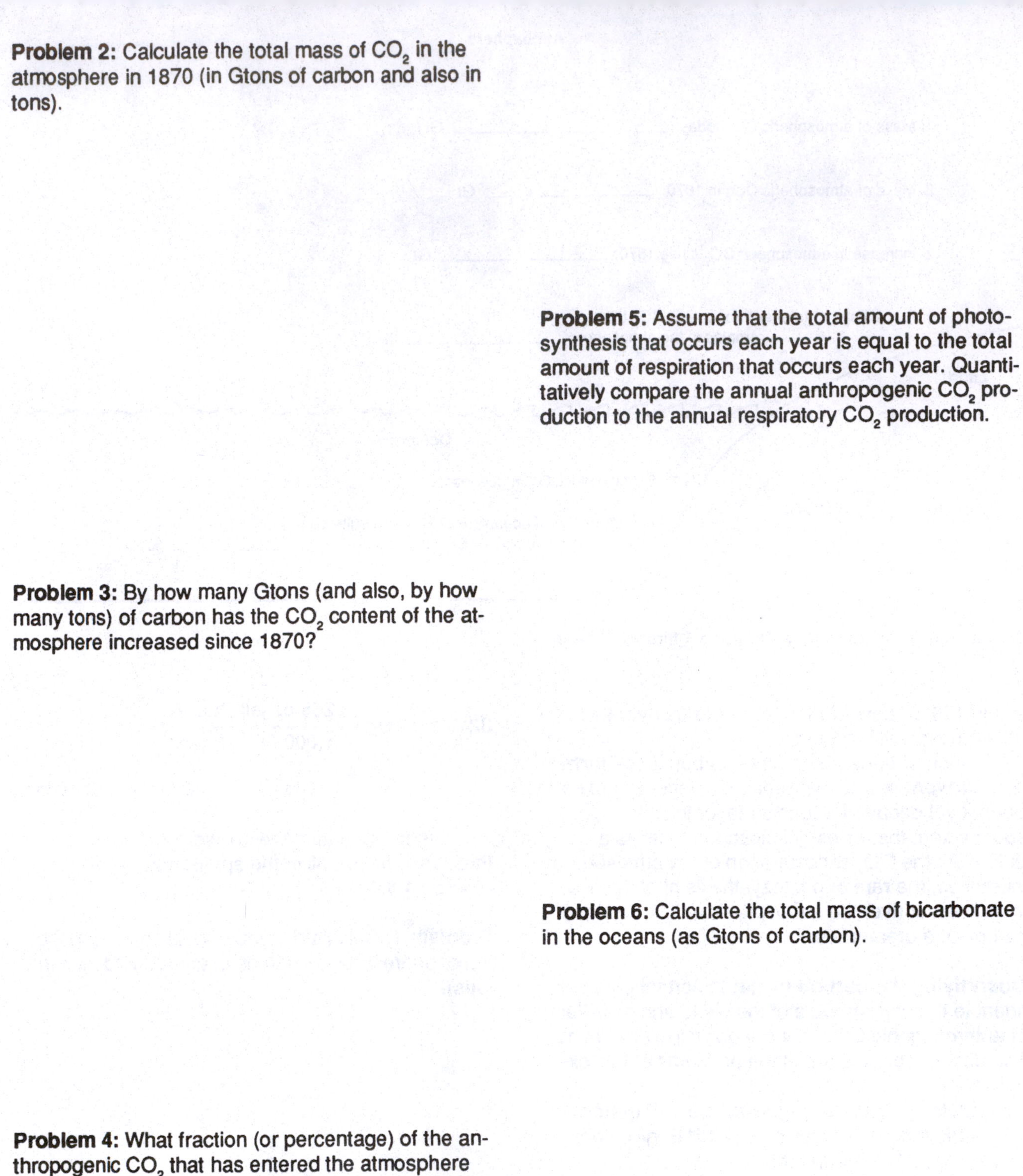

Problem 2: Calculate the total mass of CO_2 in the atmosphere in 1870 (in Gtons of carbon and also in tons).

Problem 3: By how many Gtons (and also, by how many tons) of carbon has the CO_2 content of the atmosphere increased since 1870?

Problem 4: What fraction (or percentage) of the anthropogenic CO_2 that has entered the atmosphere since 1870 still remains in the atmosphere?

Problem 5: Assume that the total amount of photosynthesis that occurs each year is equal to the total amount of respiration that occurs each year. Quantitatively compare the annual anthropogenic CO_2 production to the annual respiratory CO_2 production.

Problem 6: Calculate the total mass of bicarbonate in the oceans (as Gtons of carbon).

Problem 7: Assume that the fraction of anthropogenic CO_2 that does not remain in the atmosphere has gone into the oceans to form bicarbonate. What fraction of the total bicarbonate of the oceans is that?

Problem 8: Given your answer to Problem 7, to what extent do you think that the dissolution of anthropogenic CO_2 in the oceans has affected the ability of the oceans to respond to further additions of anthropogenic CO_2 to the oceans? Explain your reasoning.

Name: ______________________ Date: ______________ Section: ____________

Exercise 15
CLASSIFICATION OF MARINE ORGANISMS

Purpose: Life on earth is very diverse. The scientific literature contains descriptions of as many as 1.2 million species of animals, perhaps 200,000 of these living in the oceans. Biologists have estimated that a much greater number of species may be alive but still undescribed. Understanding the relationships among all of these organisms, and between the organisms and the environments in which they live and of which they form a part, requires some framework in which to organize information. Even simply cataloging the organisms would be a daunting task without some scheme of classification. As a result, scientists have developed a number of approaches to categorizing or classifying organisms and information about them. The most familiar of these is probably the taxonomic scheme of Linnaeus (kingdom, phylum, . . . genus, species). In this exercise we will examine some approaches to the classification of organisms. As you go through this exercise, keep in mind that classification schemes are human creations, designed to describe the way in which nature works. Different classification schemes focus on different features of organisms and their relationships to the communities in which they live. The best classification scheme to use depends upon the purpose to which the information being classified is to be put.

Taxonomy—the Linnaean classification scheme: In the middle of the 18th century the Swedish scientist Karl von Linné (Carolus Linnaeus in the Latinized version) devised a scheme for classifying organisms according to the closeness of the evolutionary relationships perceived among them. Under this scheme, each organism belongs to a kingdom (the most general or fundamental division into which organisms are categorized), and each kingdom is divided and subdivided into increasingly specific categories:

Kingdom
Phylum
Class
Order
Family
Genus
Species

The "scientific name" of a species includes its genus and species. The first letter of the ***generic name*** is always capitalized. Both genus and species names typically have Greek or Latin roots and are written in italics or underlined. For example, humans are *Homo sapiens* (man the wise). A scientist who describes in the scientific literature a previously unnamed organism may assign a name to that organism. (With estimated millions of kinds of organisms yet to be discovered, ample opportunity remains to assign taxonomic names.) Difficulties can arise, of course, if two scientists, unbeknownst to one another, assign different names to the same organism or the same name to different organisms.

One of the utilities of the taxonomic scheme of classification is that it provides a family tree for an organism, showing its relationship to other organisms. When fossil organisms are also included the taxonomic scheme is a useful one for illustrating the way in which newer organisms have evolved from older ones. It is a ***phylogenetic*** scheme.

While the taxonomic scheme calls for organisms to be pigeonholed, or assigned neatly to one category or another, this does not always match the way in which nature works. This is one of the shortcomings of the taxonomic scheme. For example, if we were to consider, as those who are not biologists commonly do, that all organisms were members of either the plant kingdom or the animal kingdom, we would encounter difficulties classifying a group of single-celled marine organisms called dinoflagellates (named for their two *flagella* or whiplike attachments to the cell body). In many respects, these organisms satisfy what you would probably consider to be obvious criteria for classifying them as animals. For example, they can move their flagella to provide locomotion. Many dinoflagellates are ***heterotrophic*** (i.e., they obtain their nourishment by consuming food available in their surroundings), and thus satisfy that criterion for classification as animals. But other dinoflagellates are ***autotrophic***. That is, they contain chlorophyll and can generate their own food through the process of *photosynthesis* (Exercise 14). They would, in that sense, satisfy most people's definition of a plant. However, if we did classify these autotrophic dinoflagellates as plants and the heterotrophic di-

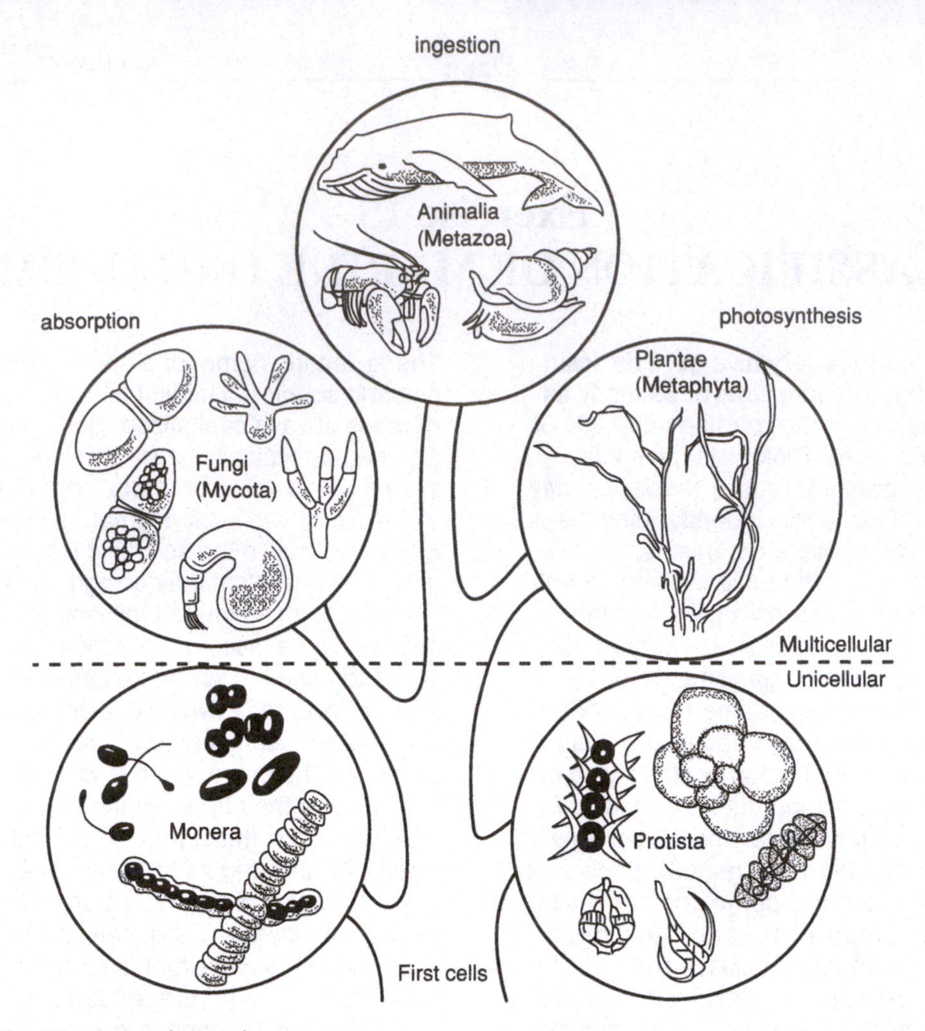

Figure 15.1 The five kingdoms of organisms.

noflagellates as animals, we would be in the unsatisfactory situation of classifying two closely related forms as belonging to two different kingdoms. The problem is that in nature things do not always fit as nicely as we might like into preconceived pigeonholes. The solution, in this case, is to redesign the pigeonholes. There is nothing inherent in the organization of life that requires that it be divided into only two kingdoms. Biologists now commonly group life forms into five kingdoms (Figure 15.1), but even that number is somewhat arbitrary and is a matter of convenience.

Of the five kingdoms into which life is commonly divided, the ***Monera*** constitute the simplest, most primitive of the kingdoms. They are single-celled, but the cell does not have a nucleus. The nuclear material (including DNA, which carries genetic information) is distributed throughout the cell. This kingdom includes the cyanobacteria (blue-green algae, which are autotrophic—they photosynthesize) as well as the bacteria (which are heterotrophic).

The ***Protista*** are also single-celled, but they are more complex than the Monera. The cell of a protist has a nucleus containing genetic material (DNA) that is surrounded by a membrane. Included in this kingdom are simple autotrophic organisms such as most algae (which you might think of as primitive plants) and heterotrophic organisms (which you might think of as primitive animals).

The kingdom of the ***Fungi*** (or ***Mycota***) consists of multicellular organisms that have many characteristics typical of plants, but they are heterotrophic. They do not photosynthesize, but like animals, they get their nutrition by incorporating organic matter from the environment. The Mycota are not abundant in the marine environment.

Multicellular plants belong to the kingdom ***Plantae*** (or ***Metaphyta***). These are not important in the ocean except in coastal waters. Multicellular animals belong to the kingdom ***Animalia*** (or ***Metazoa***).

An organism is assigned a taxonomic classification based upon its resemblance to other organisms, and its relationship to them in an evolutionary or phylogenetic sense. Resemblances among organisms may be in external form, internal functioning, or both. (Closely related organisms may not appear similar in their adult or mature stages, however. It is common for a marine organism to undergo enormous changes of form during the course of its life. Organisms that appear very different from one another in one life stage may be very similar to one another in a different life stage.)

An example of the assignment of organisms to an appropriate taxonomic classification is illustrated, using the example of some inanimate objects, in Figure 15.2a. While examples of this sort are useful in understanding how organisms can be classified taxonomically, it is important to keep in mind that these examples have the shortcoming of ignoring evolutionary or phylogenetic relationships.

A possible taxonomy of the objects in Figure 15.2a is illustrated in Figure 15.2b. To simplify this example, the only levels of categorization used are kingdom, phylum, genus, and species. Note that the objects are grouped according to degree of similarity, which in the case of living organisms, one might take to be degree of closeness in an evolutionary sense.

Problem 1: Twenty inanimate objects are shown in Figure 15.3a. Using the example given in Figure 15.2 as a model, treat these objects as if they were animate, and create a taxonomy for them. The taxonomy should be based solely on the similarities among the objects. It should ignore the uses to which the objects are put. For example, it is not relevant that objects 13 and 14 are both used in sewing (although similarities or differences in their forms should be critical to your taxonomy).

View each of the objects as fully formed, i.e., adult. Place objects for which you see no relationship whatsoever in separate kingdoms. Create as many kingdoms as you like, but remember that a taxonomy in which every object is in a different kingdom is of no use. As in the example in Figure 15.2, subdivide each kingdom according to phylum, genus, and species. Each of the 20 objects is a different species.

Sketch your taxonomy in Figure 15.3b. *Note that there is no single right answer to this question, but that the answer you give must make sense and be internally consistent.*

Problem 2: Some of the characteristics of several phyla of important marine animals are listed in Table 15.1. Drawings of several common marine animals are shown in Figure 15.4. Using the characteristics listed in the table, determine the phylum to which each animal belongs. Write the phylum (and the common English name of the animal when you can identify it) in the space provided in Table 15.2. (Note: Your instructor may ask you to deter-

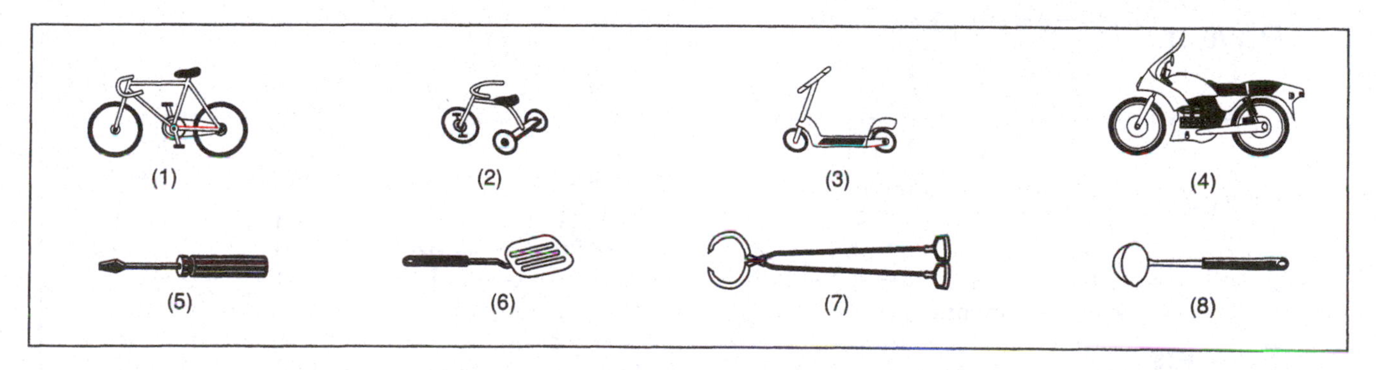

Figure 15.2a. A series of objects used in an example of taxonomic classification (see text).

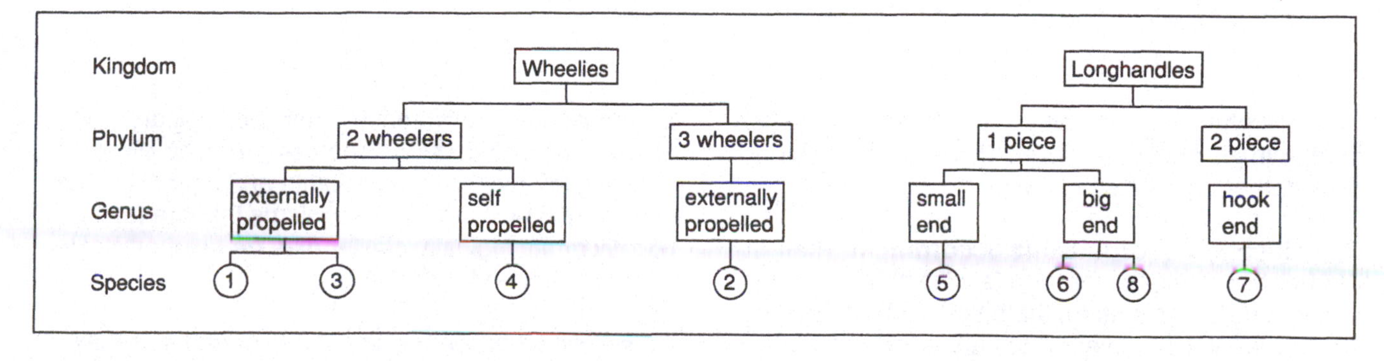

Figure 15.2b. A taxonomy of the objects shown in Figure 15.2a.

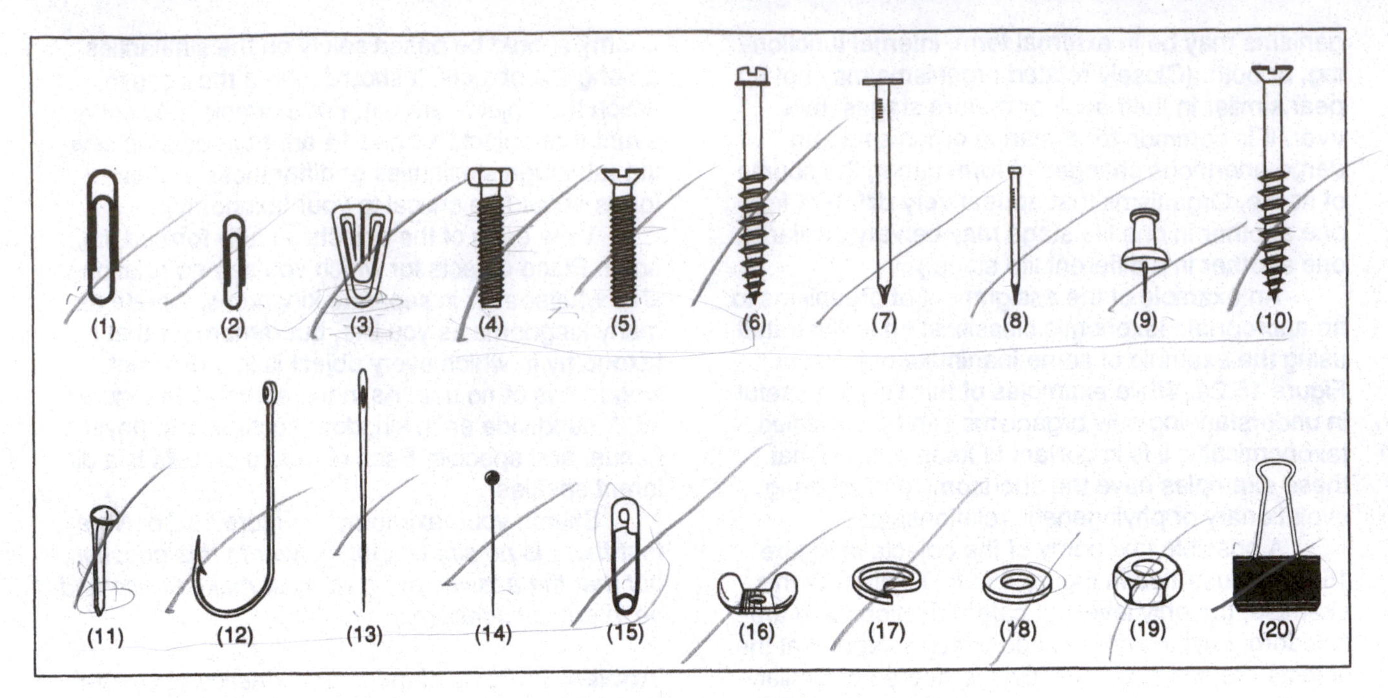

Figure 15.3a. A series of objects to be classified taxonomically in Problem 1.

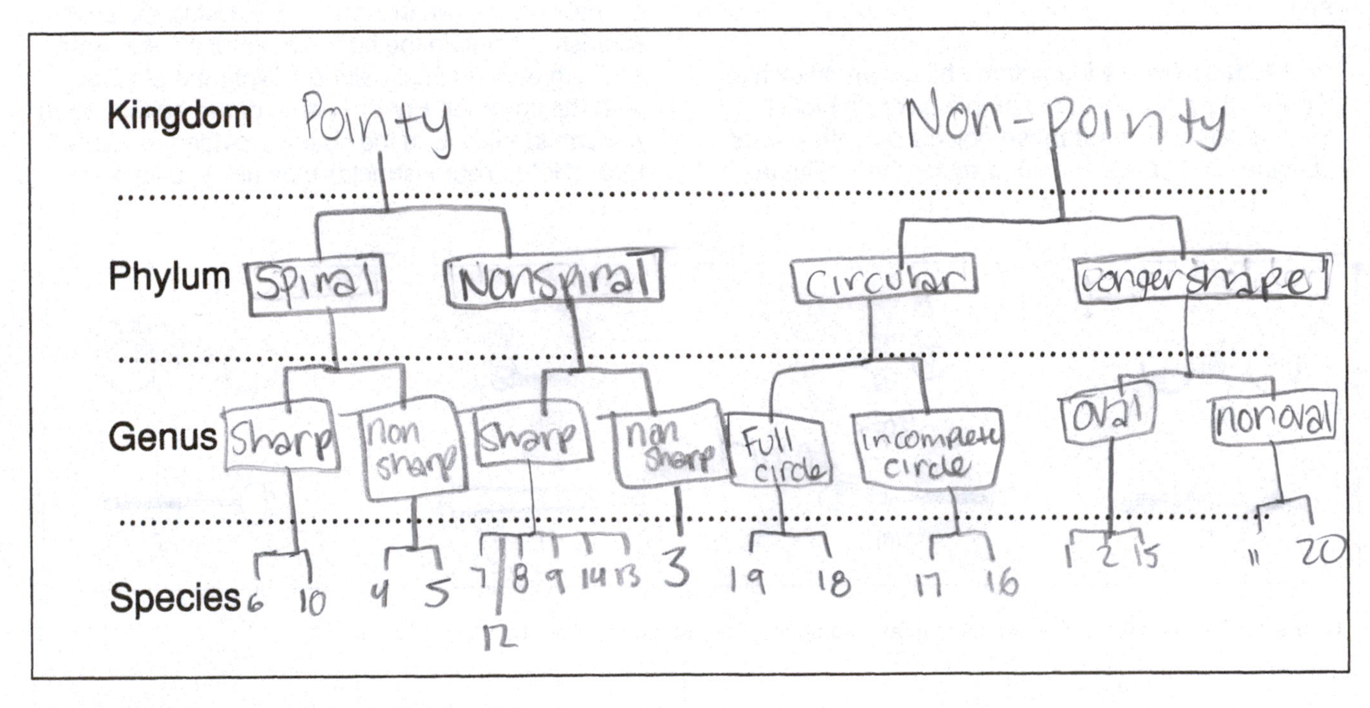

Figure 15.3b. Put your answer to Problem 1 here.

mine the phylum of preserved biological specimens, or of animals shown on slides, filmstrip, or videotape, rather than those of Figure 15.4.)

Classification of organisms according to their life habits: While taxonomic classification is useful in grouping organisms on the basis of phylogenetic similarities, it is not useful if we are concerned with how organisms are related to their environment or how they interact with each other. Ecological (or functional) classification schemes are useful for organizing information about plants and animals in the context of how they live. Organisms that are functionally or ecologically similar may be taxonomically very different and vice versa.

Classification according to habitat. Organisms may inhabit the water column (in which case they are

Table 15.1. Characteristics of some important marine phyla.

Phylum	*Characteristics*
Porifera	Sponges. Collections of cells, not differentiated into tissues. Filter feeders. No hard parts except spicules (needle-like parts) distributed through the organism.
Coelenterata (or Cnidaria)	Jellyfish, corals, etc. Radial symmetry. Two-cell layered body wall with a single opening to cavity. Stinging cells.
Ctenophera	Comb jellies. Eight-sided radial symmetry. No stinging cells.
Platyhelminthes	Flatworms. Bilateral symmetry.
Nematoda	Roundworms. Marine forms are mostly free-living and benthic. Usually 1 to 3 mm long.
Bryozoa (or Ectoprocta)	Moss animals. Benthic colonies that branch or encrust. Have a lophophore (a horseshoe-shaped feeding structure with ciliated tentacles).
Mollusca	Clams, snails, squid, octopuses, etc. Soft bodies with a muscular foot. Many varieties have a mantle that secretes a calcium carbonate shell.
Annelida	Segmented coelomate (i.e., hollow-bodied) worms. Musculature, nervous, excretory, and reproductive systems, may be repeated in many segments. Mostly benthic.
Arthropoda	Crabs, lobsters, shrimp, copepods, etc. Jointed-legged animals with a segmented body covered by a hard exoskeleton.
Echinodermata	Starfish, sand dollars, sea urchins, sea lilies, etc. Spiny-skinned animals with radial symmetry (usually five-fold). Water vascular system.
Chordata	Notochord or (in vertebrates) spinal cord. Most examples are vertebrates, including primitive fish, bony fish, amphibians, reptiles, birds and mammals. (Humans are chordates.)

called ***pelagic***) or may inhabit the bottom of the ocean (in which case they are called ***benthic***). Benthic fauna may be further classified according to whether they are ***epifaunal*** (live on the surface of the ocean bottom) or ***infaunal*** (burrow into the ocean bottom).

Classification according to mobility (or motility). Many pelagic organisms, both plants and animals, have only very limited means to control their location in the water column. These organisms depend upon the currents to move them. They are called ***plankton***. Autotrophs of this type are ***phytoplankton*** and heterotrophs are ***zooplankton***. Pelagic organisms that can swim are called ***nekton***. Benthic organisms may be attached to the bottom (***sessile***) or may be able to move about the bottom (***motile***). Some organisms normally live on the bottom but have the ability to swim above the bottom. These are ***nektobenthos***. It is often possible to classify an organism according to its mobility simply on the basis of its appearance.

Problem 3: Several marine organisms are depicted in Figure 15.5. In Table 15.3, classify each of these organisms according to its motility and write a few words explaining why you assigned the classification that you did.

Classification according to mode of nutrition. Plants are autotrophic. They make their own food. Animals are heterotrophic. They must obtain their nutrition from outside their bodies. There are many different ways in which animals do this. ***Suspension feeders*** have mechanisms for filtering out the tiny particles of food (for example, plankton) that are sus-

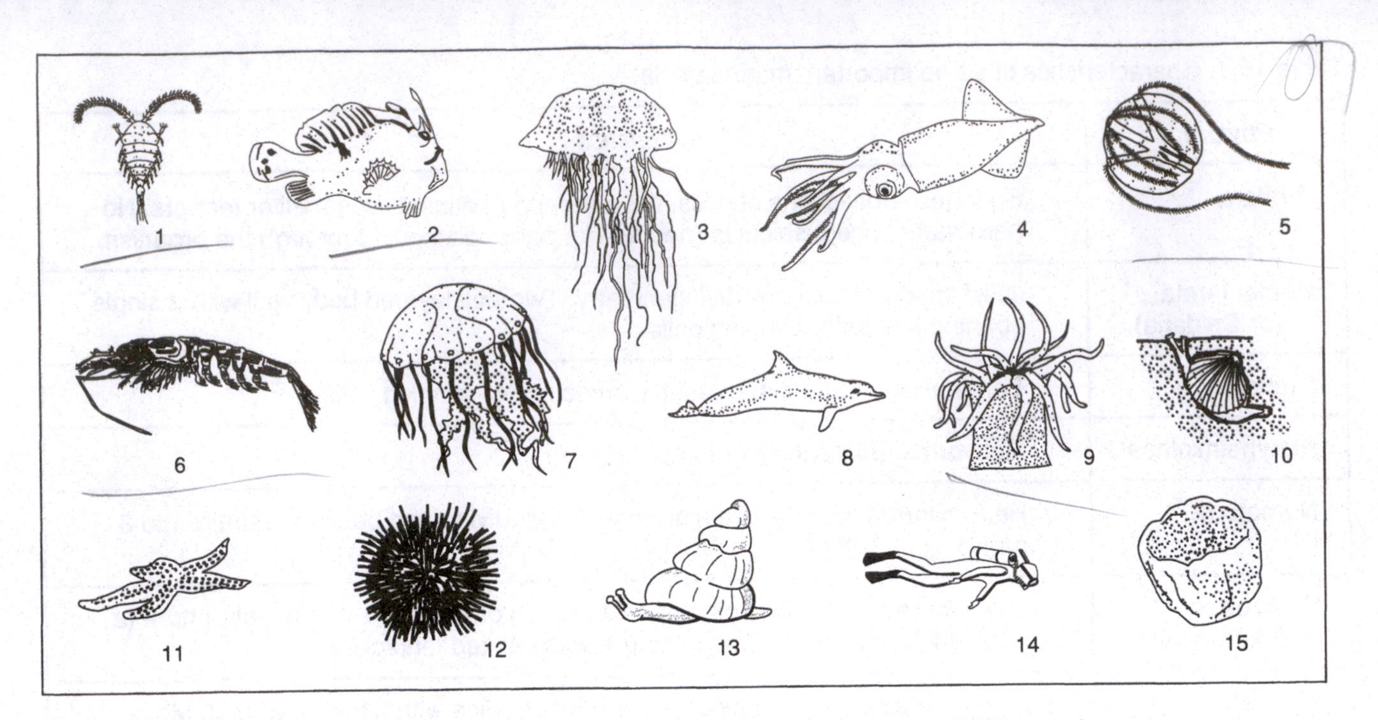

Figure 15.4. Pictures of organisms for use in Problem 2. Note that the organisms are of different sizes.

Table 15.2. Write answers to Problem 2 here.

Picture or specimen number	*Phylum*	*Common name*
1	Arthropoda	CorePod
2	Chordata	fish
3	Cnidaria	Jellyfish
4	Mollusca	Squid
5	Ctenophera	comb jelly
6	Arthropoda	Shrimp
7	Cnidaria	Jellyfish
8	Chordata	Dolphin
9	Cnidaria	Coral
10	Mollusca	Snail
11	Echinodermata	Star fish
12	Echinodermata	Sea urchin
13	Mollusca	Snail
14	Chordata	Human
15	Porifera	Sponge

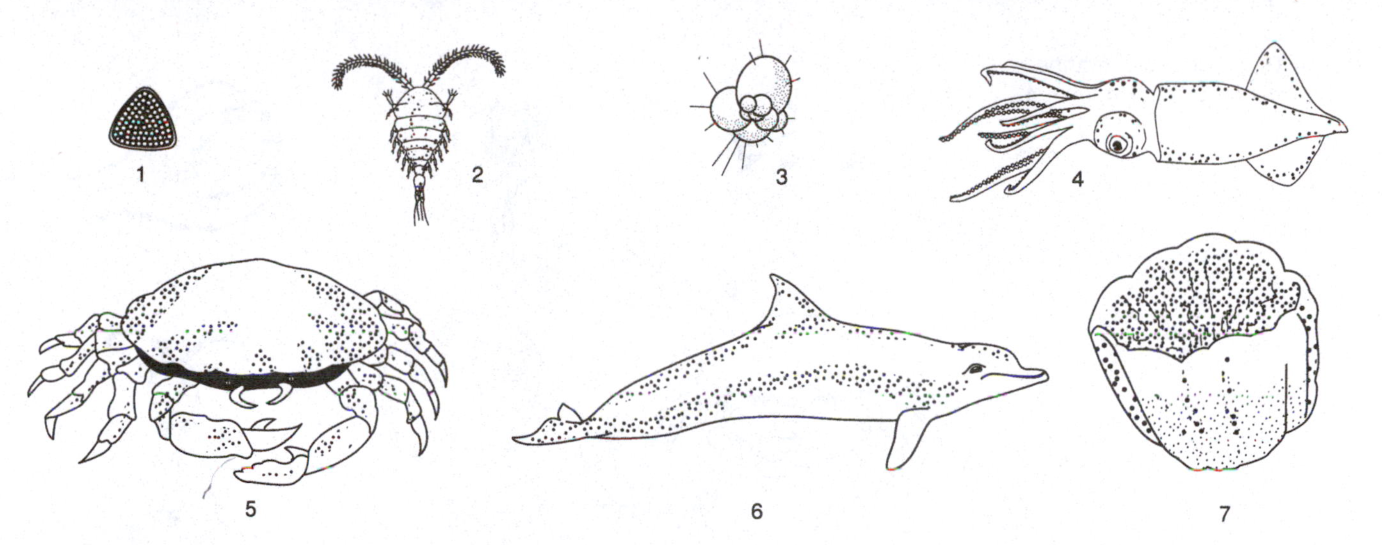

Figure 15.5. Pictures of marine organisms for use in Problem 3. Organisms 1 through 3 are smaller than 1 cm. Organisms 3 through 7 are larger than 1 cm.

Table 15.3. Write answers to Problem 3 here.

Picture No.	*Motility*	*Explanation*
1	Plankton	They need water to move them to get food
2	Nekton	They can swim to their food
3	Plankton	They need water to move them to get food
4	Nekton	They can swim to their food
5	Nektobenthos	They can swim up and down
6	Nekton	They can swim to their food
7	Benthic	They stay attached to the bottom

pended in sea water everywhere. There are many varieties of mechanisms that serve this purpose. Examples include arrangements of hair-like cilia that can trap food particles and direct them to an animal's mouth, and the large comb-like baleen of the baleen whale. Some suspension feeders, the mucoid filter feeders, use a layer of mucus to trap suspended food particles. ***Deposit feeders*** eat mud or sand, digesting some of the organic material in the sediment and excreting the bulk of the material that they cannot use. Herbivores eat large algae. ***Predators*** attack and catch prey. Predators may often be recognized because they possess equipment that aids them in this task, for example the stinging cells of the jellyfish, the crushing claws of the American lobster, and the teeth of the shark.

Problem 4: Several marine organisms are depicted in Figure 15.6. In Table 15.4, classify each of these organisms according to its mode of nutrition and write a few words explaining why you assigned the classification that you did.

Problem 5: Examine the pictures of the clam in Figure 15.7 and the barracuda in Figure 15.8. Based upon what you can see in each picture, classify this

Figure 15.6 Pictures of marine organisms for use in Problem 4.

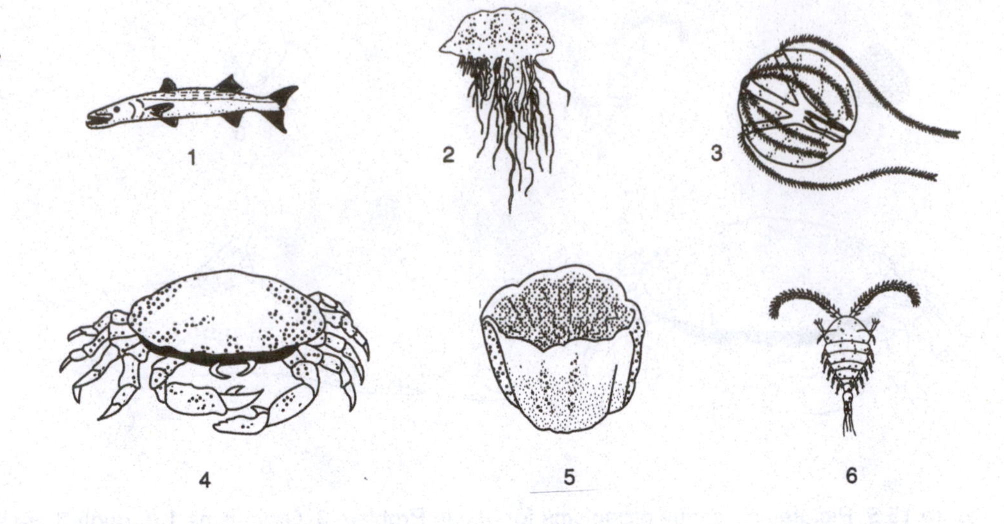

Table 15.4. Write answers to Problem 4 here.

Picture No.	*Mode of nutrition*	*Explanation*
1		
2		
3	Suspension feeding	
4	Predator	
5		
6		

clam and the barracuda in terms of habitat, motility, and means of nutrition. Explain why you came to the conclusions you did.

Clam:

Clams are nektobenthos because they can move from sand to water. they are heterotrophic predators that search for their food They are typilly living in the sand burrowing they're infaunal

Barracuda:

This animal lives in the Pelagic zone b/c it lives in the middle of the water. It's a heterotrophic predator b/c it goes to hunt it's food. And it's motility is Nekton b/c it can swim anywhere to catch it's food

Figure 15.7. A series of sketches showing a clam burrowing into sediment.

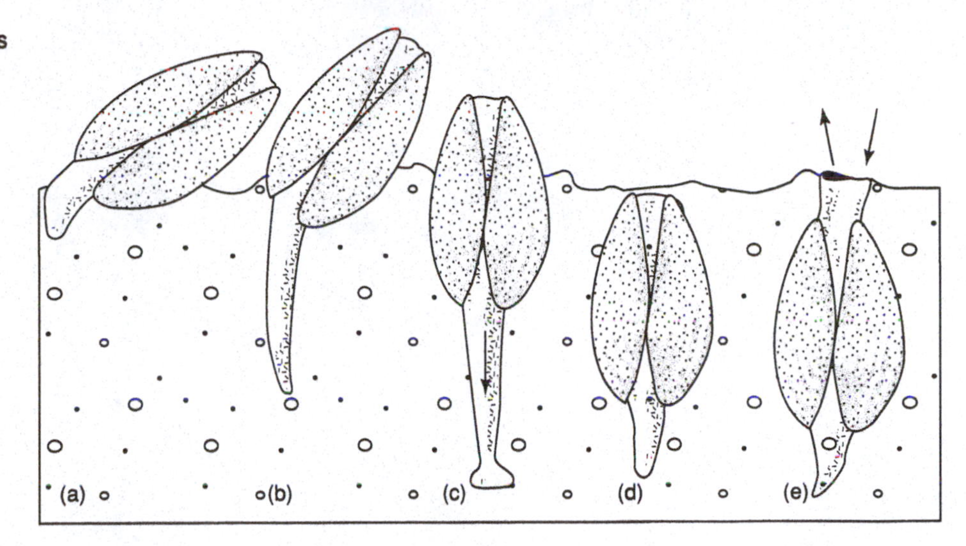

Figure 15.8. The barracuda.

Name: ______________________ Date: ____________ Section: __________

Exercise 16
FOOD CHAINS, FOOD WEBS, AND BIOLOGICAL PRODUCTIVITY

Purpose: The rate at which biological materials are produced in the oceans is an important determinant of the capability of the oceans to feed the world's population. Biological activity plays a critical role in the control of the chemistry of the atmosphere. As discussed more fully in Exercise 14, the marine biosphere is an important agent of the consumption and production of both atmospheric oxygen (O_2) and carbon dioxide (CO_2). In this exercise you will become familiar with the factors that control the amount of biological material produced in the oceans and the way in which biological material and energy flow through the biosphere.

Energy flow in the biosphere: Biological materials consist largely of chemical compounds of carbon, hydrogen and oxygen, with smaller amounts of the ***nutrients*** nitrogen, phosphorous, silicon (in the case of some marine plants and animals), and a host of ***trace elements*** in much smaller concentrations. These compounds are called ***organic matter***. Most organic matter is unstable in the presence of atmospheric oxygen. It becomes oxidized, yielding H_2O, CO_2, and ***energy*** (and small amounts of the nutrients and trace elements) as oxidation products.

The fact that organic matter is spontaneously oxidized in the presence of the atmosphere, and that the oxidation process yields energy, tells us that organic matter (or biological material) cannot form without the addition of energy. Plants are the ***primary producers*** of organic matter, through the process of photosynthesis.

$$6CO_2 + 6H_2O + \text{energy} \longrightarrow C_6H_{12}O_6 + 6O_2$$

The energy which goes into the process is, in almost every case,[1] solar energy (sunlight). The primary product, $C_6H_{12}O_6$, is glucose, a sugar. The glucose is converted by plants into a host of compounds that make up the plant. Plants are eaten by herbivores, the herbivores by carnivores (all of which are grouped together as ***consumers***), and so on, up through the food chain. At every step of the way, some of the organic matter is constructed into new, complex compounds that the consumers use to build their bodies or to function. Another fraction of the organic matter is broken down into simpler compounds (ultimately CO_2 and H_2O). This latter process is accompanied by the liberation of energy, and it is that energy that the consumers use to enable them to function (e.g., to feed, to move, to reproduce, and in the case of warm-blooded animals, to maintain body temperature). Bacteria (decomposers) perform very similar activities, breaking down organic materials, including dissolved organic matter, into the simplest of compounds. The first law of thermodynamics (the law of conservation of energy) describes a limit on the amount of energy that an organism can obtain by use of the organic matter. *No more energy can be obtained by all of the consumers and decomposers through which organic matter passes than was captured (as sunlight) by plants, the primary producers.* The second law of thermodynamics tells us that *even less* energy is available to the consumers and decomposers than was captured by the producers. Accompanying each chemical process, some of the energy is lost to the environment as heat. It is not available for functioning.

Problem 1: The input of solar energy to the surface of the earth varies a great deal as a function of latitude, season of the year, and time of day. Averaged over the entire globe and over the entire year, the input of solar energy to the surface of the earth is approximately 2 million calories per square meter per day [2×10^6 calories/(m^2 day)]. A typical adult human requires approximately 2×10^6 calories per day for normal functioning.[2] Assume that you were

---

[1] In recent years it has been discovered that bacteria in some benthic environments are capable of using the energy of chemical bonds of substances in the environment (rather than light energy) to construct the basic constituents from which their cells are built. This is called *chemosynthesis*. Environments in which chemosynthesis has been found to be important include sulfur-rich submarine hot springs and natural gas and oil seeps.

[2] A calorie is the amount of energy required to heat 1 gram of H_2O (i.e., approximately 1 cubic centimeter) 1 degree Celsius. The calorie used in dietetics is equal to 1000 calories as defined above. Thus, a typical adult calorie requirement of 2,000 calories per day is equivalent to 2,000,000 (or 2×10^6) calories per day as defined above.

able to get all of your energy (i.e., food) requirements from plants, and that plants converted 100 percent of the solar energy striking the surface of the earth into food energy. Calculate the area of the earth's surface that would be required to provide your food requirements.

The assumptions in the calculation above are clearly unrealistic. Reconsider the problem but assume the following: a) that only one percent of the solar energy striking the surface of the earth is captured by plants and used for their biological processes (including increasing their biomass); b) that only 20 percent of the energy captured by the plants is used for the formation of biomass (and hence is available for consumption) while 80 percent is used up in other life processes of the plant; c) that in competition with other consumers, humans are able to consume only 1 percent of the primary production and non-humans (chiefly animals, fungi, and bacteria) consume 99 percent. Now calculate the area of the earth's surface that would be required to supply your daily needs of 2×10^6 calories.

Even these assumptions are not realistic, because much of the area of the land, and even more of the area of the oceans, is not covered by plants.

The area of the earth (continents and oceans together) is 5.1×10^{14} m^2. Using the answer you obtained immediately above, and ignoring distinctions between ocean and dry land, calculate the number of people that the solar energy input might support on the planet earth.

Note that the answer you calculated above required a large number of assumptions, none of which may describe the functioning of the biosphere very accurately. Nevertheless, the problem should serve to illustrate one of the constraints that exists on the ability of the earth to support an ever-growing population.

Nutrients and primary production: Even with abundant sunlight, primary production remains low when plants do not have sufficient nutrients. Critical nutrients in marine systems are nitrogen (present as nitrate, NO_3^-, nitrite, NO_2^-, and ammonium, NH_4^+) and phosphorous (present as phosphate, PO_4^{-3}).

The decreasing availability of light with increasing water depth limits plant growth to the ***euphotic zone***, typically the upper hundred meters or less of the water column. As plants grow, they are very efficient at incorporating nutrients from the water. Commonly most of the nitrate and phosphate in the near-surface zone may be in the biomass rather than in solution. As a result of mixing and sinking under the force of gravity, there is typically a downward movement of plants and the animals dependent on the plants, as well as animal wastes, and dead animal and plant remains. With downward migration, respiration by plants and animals and bacterial decomposition gradually return nutrients to the water as dissolved constituents. These processes, taken together, result in the removal of nutrients from the surface layers of the water and their concentration in deeper waters. The greatest concentration of nutrients typically occurs at depths of several hundred meters. This downward movement is referred to as the biological pumping of nutrients.

Remembering that biological productivity in the presence of sufficient sunlight is limited by the availability of nutrients, it is not surprising that the regions of highest productivity are those in which nutrient-rich waters are brought to the surface from depths of several hundred meters. This may be the result of upwelling or the seasonal weakening of the thermal stratification of the water column, allowing vertical mixing to occur.

High latitudes: In polar regions, sunlight can sometimes be the limiting factor in biological productivity. During the winter the sun is low in the sky, and above the Arctic (or Antarctic) Circle it may not rise for a period of days, weeks or even months at time. In the Arctic Ocean, ice cover filters out much of the light before it can penetrate the water. On the other hand, during the summer the amount of sunlight reaching the surface can be very high, and the absence of strongly developed vertical density stratification can permit vertical mixing and the enrichment of surface waters in nutrients.

Tropical latitudes: In these regions there is typically a strongly developed thermocline and nutrients are commonly depleted from the upper mixed layer, limiting biological productivity. Exceptions to this occur in regions in which there is upwelling or in which nutrients are locally brought into the neritic zone by runoff from the continents.

Mid-latitudes: Either sunlight or nutrient availability can be important in determining biological productivity. Nutrient availability can be enhanced by upwelling, continental runoff, or wintertime vertical mixing as the vertical density contrast diminishes.

Problem 2: In mid-latitudes it is common for biological productivity to be higher in the spring and the fall than it is either in the winter or the summer. Why is this so? (Hint: Consider separately the factors that are important in limiting productivity in each season.)

____

____

____

____

____

____

____

____

____

____

Problem 3: Experiments have shown that phytoplankton in the north Pacific and in Antarctic waters increase their production when iron is added to the nutrients that exist naturally in the environments. This suggests the possibility that iron may be the limiting nutrient in these waters. It has given rise to the suggestion that polar waters might be artificially fertilized with iron as a way to reduce atmospheric CO_2 levels and avert greenhouse warming. Explain how this might work.

____

____

____

____

____

____

____

____

____

Relationship between dissolved oxygen and nutrient concentration: Figure 16.1 shows a typical profile of nutrient concentration and oxygen concentration for a typical tropical or mid-latitude location. Note that the concentration of dissolved oxygen drops off markedly in the upper thousand meters of the water column, reaching quite low values in the oxygen minimum zone. Over roughly the same range of depths the nutrient content increases, reaching a maximum at depths similar to those of the oxygen minimum.

Problem 4: Explain the cause of the complementary relationships seen between the oxygen concentration and the nutrient concentration in Figure 16.1.

The biogeochemical cycle of nitrogen: The nitrogen cycle is illustrated schematically in Figure 16.2. Although nitrogen is the most abundant gas in the atmosphere (molecular N_2) and is dissolved as N_2 in the oceans, most plants are not able to use nitrogen in this form. A group of bacteria, the ***nitrogen-fixing bacteria***, are, however, able to take N_2 and convert it into nitrate or nitrite, forms which most plants are able to use.

Problem 5: The nitrogen cycle depicted in Figure 16.2 is complex but important. As an aid in understanding it, describe in words the way nitrogen moves through the biogeochemical cycle.

Trophic levels and biomass pyramids: Of the organic matter produced by plants, a large fraction is oxidized by the plants themselves. Some plants and parts of plants then die and are decomposed by

Figure 16.1. Oxygen concentration (solid curve) and nutrient concentration (dotted curve) in seawater as a function of depth in a typical tropical or mid-latitude location.

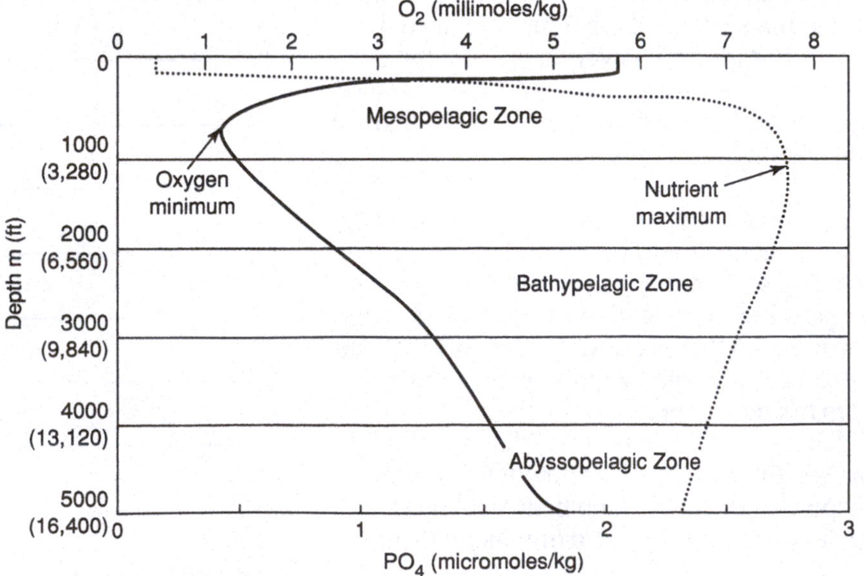

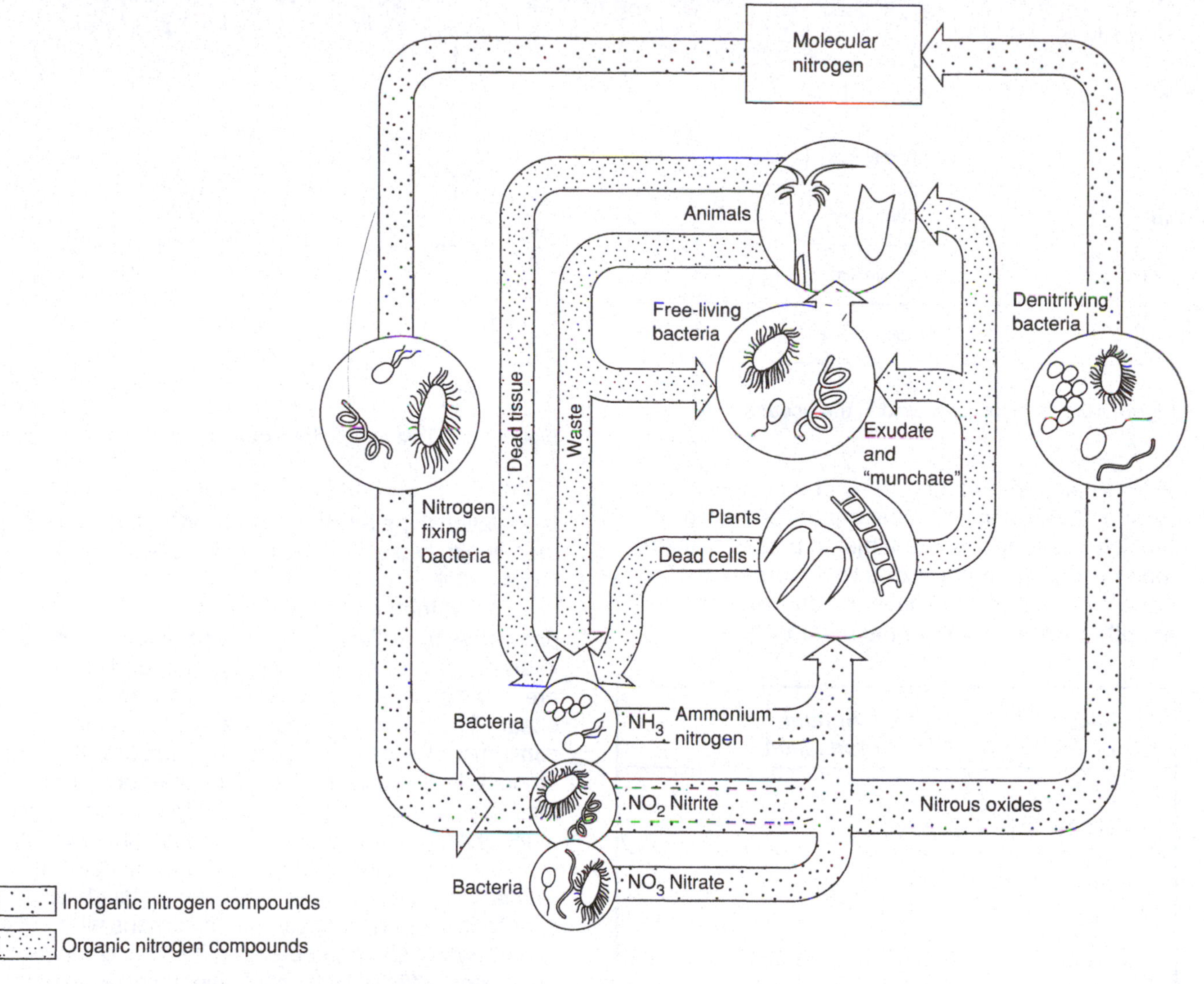

Figure 16.2. The nitrogen cycle. Denitrifying bacteria turn nitrate, nitrite, and ammonia into N_2, a form of nitrogen that is inaccessible to plants. Nitrogen-fixing bacteria change N_2 back into nitrate and other accessible forms of nitrogen. Exudate refers to ***cytoplasm*** (cellular material) that is lost directly to seawater as phytoplankton age. Munchate refers to cytoplasm that is spilled into the ocean as they are eaten by zooplankton.

bacteria while another fraction of the plant matter is eaten by herbivores. These excrete much of the plant matter (which they are not able to use) and much of that is in turn decomposed by bacteria. Of the food that is not excreted, some goes to the growth of the herbivore, some provides energy for its locomotion, for its feeding process, etc., and some provides energy for reproduction. Typically the herbivores use up or excrete about 90 percent of the organic matter they eat and incorporate something like 10 percent into their bodies. That 10 percent is available to be eaten by ***primary carnivores*** (carnivores that eat herbivores) who also divide their food among excrement, growth, locomotion, reproduction, etc. Again, something like 10 percent of the primary carnivore's diet is incorporated into its body, available for eating by ***secondary carnivores*** (carnivores that eat primary carnivores), and so on.

We can illustrate this general scheme by a food pyramid, as in Figure 16.3. Each layer of the pyramid is called a ***trophic level***. The primary producers occupy the first trophic level, the herbivores the second trophic level and so on.[3]

Problem 6: Consider the food pyramid in Figure 16.3. Assume that each trophic level operates at an efficiency of 10 percent. That is, 10 percent of the food eaten by that trophic level is incorporated into the organisms and 90 percent is excreted or pro-

[3] Some authors place the primary producers on the zeroth trophic level, the herbivores on the first trophic level, etc.

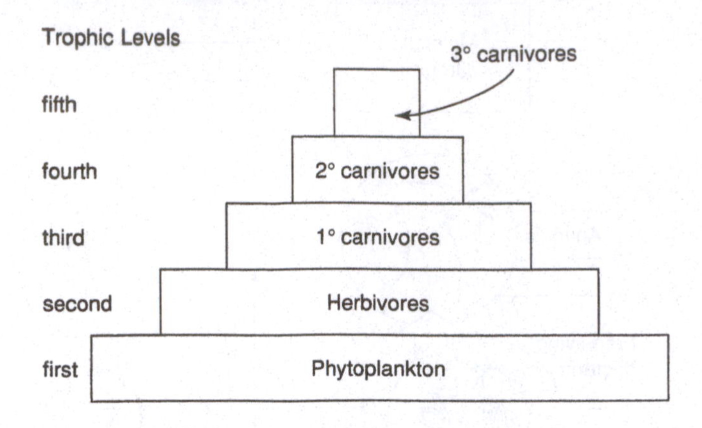

Figure 16.3. The food pyramid in the oceans.

vides energy for the organisms of that level. Consider a situation in which the annual primary production in a region is 1,000 metric tonnes (1 metric tonne = 1,000 kilograms or a little more than 2,000 lbs.). What will be the annual production at each of the other trophic levels of the pyramid ?

Trophic Level	*Annual Production*
sixth	____________
fifth	____________
fourth	____________
third	____________
second	____________
first	1,000 metric tonnes

It is very important, when considering the fish production of the world, that fish in upwelling zones are very frequently herbivores or primary carnivores, that fish in neritic regions (i.e., the waters overlying the continental shelves) are often tertiary carnivores, and that fish in the open oceans often occupy the sixth trophic level. It is apparent from your answers to Problem 6 that the shorter a food chain is, the greater is the amount of food available to the ultimate carnivore.

Food chains and food webs: Some organisms occupy fairly simple positions in the trophic scheme of things in that their food consists of only a single organism, and they are in turn eaten only by a single organism. In such cases it is convenient to describe the relationships among organisms by means of a food chain, two examples of which are shown in Figure 16.4.

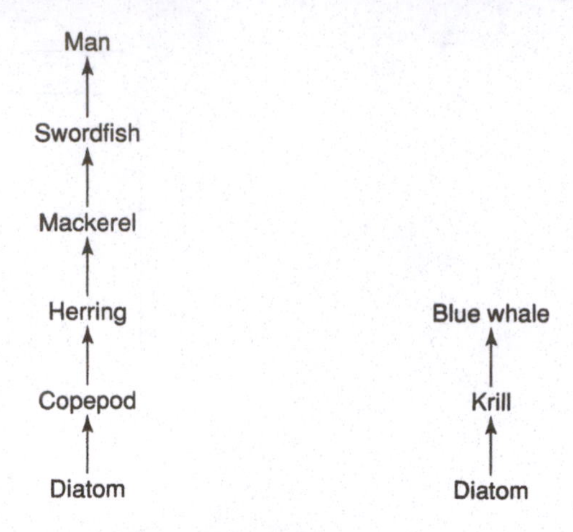

Figure 16.4. Two simple food chains.

Short food chains are characteristic of highly productive upwelling environments, such as those in areas of coastal upwelling. Longer food chains are common in less productive regions of the ocean. The amount of energy available to the final consumer of a long food chain is much less than the amount available to the final consumer of a short food chain. That is, the longer the food chain is, the more energy is lost. Using the two examples in Figure 16.4, the same amount of primary production (or diatoms) is capable of supporting a much larger population of blue whales than of humans who eat a diet of swordfish. (Of course the humans could feed much more efficiently by confining their diets to the diatoms, copepods, or even herring or mackerel instead of swordfish.)

In some cases the relationships among organisms are more complex and it is more appropriate to describe them in terms of a food web, as shown in Figure 16.5.

Problem 7: An animal at the top of a food web (e.g., the North Sea herring in Figure 16.5) is in less danger from environmental changes than is an animal at the top of a simple food chain (e.g., the Newfoundland herring of Figure 16.4). Why do you think that this is so?

____

____

____

____

____

Figure 16.5. A food web.

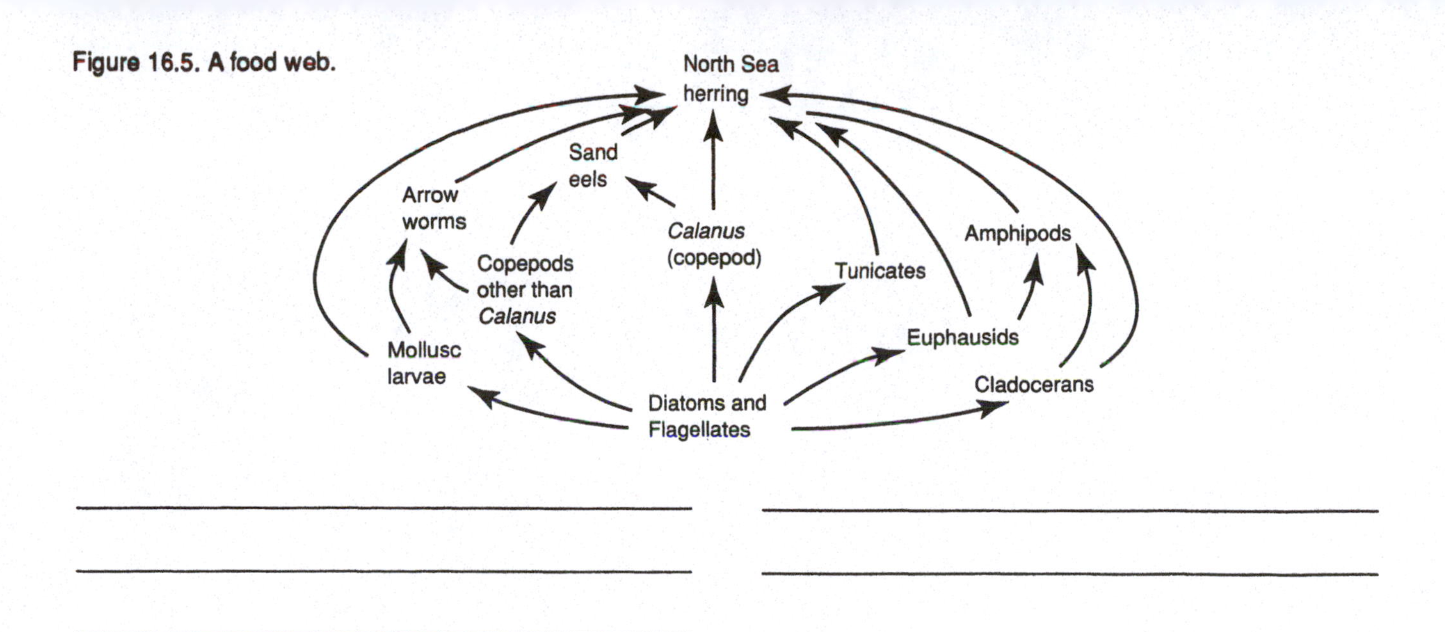

Glossary

Abyssal plain A flat depositional surface extending seaward from the continental margin.

Active margin A continental margin that is tectonically deformed because it collides with another lithospheric plate. It is the leading edge of the plate as it moves away from an oceanic spreading center.

Amorphous Characterized by a disordered arrangement of atoms in a solid. Amorphous materials may be contrasted with crystalline substances. In the latter, the atoms are arranged in three dimensions in a regular, orderly fashion.

Animalia The kingdom of many-celled animals. Term used interchangeably with *Metazoa*.

Anomaly A condition in which a measured quantity (e.g., the intensity of the magnetic field measured at a location) has a value that is greater than or smaller than an expected value.

Anthropogenic Caused by the activities of human beings.

Aragonite A mineral with the chemical composition $CaCO_3$ (calcium carbonate). Aragonite is one of two common minerals of this composition produced by marine organisms. The other is calcite.

Aseismic ridge A chain of volcanic seamounts such as the Hawaiian-Emperor seamount chain. These exhibit little seismic activity and are of different form and origin than the mid-ocean spreading ridges about which seafloor spreading occurs.

Asthenosphere A plastic layer in the upper mantle of the Earth, underlying the lithosphere. The top of the asthenosphere may be as shallow as 10 km (6.2 miles) under the oceans and is considerably deeper under the continents. Plastic movement within the asthenosphere permits both lateral and vertical motion of the lithospheric plates.

Autotrophic Describing a plant or bacterium that can synthesize organic compounds from inorganic nutrients. Autotrophs generate their own food.

Axis of the magnetic dipole A straight line drawn through the earth's north and south magnetic poles.

Backshore The inner portion of the shore, lying landward of the mean spring-tide high-water line. The backshore is acted on by the ocean only during exceptionally high tides and storms.

Backwash A thin layer of water running back down the beach. The return to the ocean or lake of that portion of the swash that does not infiltrate the beach.

Bathymetric chart A chart of a portion of the ocean that depicts the depth of the water.

Bathymetry Description of the depth of the water of the ocean or lakes.

Bathythermogram A diagram showing the variation of the temperature of seawater with depth at a chosen location.

Benthic Pertaining to the ocean bottom and the organisms that live in or on the bottom.

Bicarbonate ion The most abundant product of the dissolution of CO_2 in the ocean. HCO_3^- forms by the dissociation of carbonic acid.

Bimodal distribution A frequency distribution of observations that is characterized by two ranges of maximum values.

Biogenous Referring to components of sediment that are produced by living organisms. These components include intact and fragmented shells, coral reefs, housings of diatoms, radiolaria, and foraminifera, and particles produced by coccolithophores. Most biogenous components are calcium carbonate or opaline silica.

Biogeochemical cycle The natural cycling of compounds among the living and non-living components of an ecosystem.

Biological productivity The rate of primary production that occurs in a region. It may be limited either by the availability of sunlight or the availability or nutrients.

Biological pumping The downward movement of nutrients out of the surface waters of the ocean as the result of incorporation in plants and the sinking of plants and their remains, animals that eat plants or other animals, and their remains and wastes.

Bluff A long low cliff in the backshore region of the coastal zone.

Buffered system A chemical system that has the capacity, through internal chemical reactions, to minimize the impact of chemical changes that are imposed upon it. The ocean-atmosphere system is buffered with respect to CO_2.

Calcite A mineral with the chemical composition $CaCO_3$ (calcium carbonate). Calcite is one of two common minerals of this composition produced by marine organisms. The other is aragonite.

Carbon cycle The movement of carbon through the atmosphere, oceans, biosphere, and solid earth. The cycle incorporates both the formation and destruction of organic compounds and the relationship between oxidized and reduced carbon reservoirs.

Carbonate ion A dissociation product of bicarbonate ion. $CO_3^{=}$ is an essential building block of the calcium carbonate ($CaCO_3$) minerals calcite and aragonite.

Chart A map designed for use by mariners or a map that displays distribution of any of a variety of kinds of information, e.g., hydrographic data.

Chlorinity A measure of the salt content of seawater. Chlorinity is defined as the chloride content of seawater, expressed in grams per kilogram (g/kg) or parts per thousand ($^{\circ}/_{\circ\circ}$) by weight.

Clay Particles with sizes smaller than .004 mm (4 μm). Finer than silt.

Coastal zone The region offshore and onshore in which the rocks and sediments are affected by waves, tides, and longshore currents.

Coastline Landward limit of the effect of the highest storm waves on the shore.

Cobble Particles with sizes between 64 mm and 256 mm. Finer than boulders and coarser than pebbles.

Coccolithophore A microscopic unicellular marine plant that secretes tiny particles of calcite (calcium carbonate). These particles (coccoliths) are the predominant constituents of biogenous sediments in many areas of the ocean.

Compass rose A diagram on a map or chart that shows directional information. On a hydrographic chart the compass rose typically shows the direction of both True North and Magnetic North. It includes a circle, graduated in degrees of arc, which can be used as an aid in determining compass directions and plotting courses from one point to another.

Conic projection A map projection generated by placing a cone tangent to the surface of the globe and projecting from the center of the globe through the surface of the globe to the cone. When unrolled, the cone is a map. Distortion is least where the cone and globe are tangent and increases away from the line of tangency.

Consumer An organism that must obtain its nutrition by consuming organic matter produced by other organisms. A heterotroph.

Continental margin The zone extending from the shoreline to the deep ocean basin. It includes the continental shelf, continental slope, and continental rise.

Continental rise A gently sloping depositional surface at the base of the continental slope.

Continental shelf A gently sloping depositional surface extending from the low-water line to a point around the margin of the continent where the slope increases markedly.

Continental slope A relatively steeply sloping surface that lies seaward of the continental shelf.

Convergent boundary A lithospheric plate boundary at which adjacent plates converge, producing ocean trench–island arc systems, ocean trench–continental volcanic belt systems, or folded mountain ranges.

Coriolis Effect An effect resulting from the earth's rotation that causes moving objects to be deflected to the right in the Northern Hemisphere and to the left in the Southern Hemisphere.

Cosmogenous Referring to components of sediment that are derived from outer space, e.g., micrometeorites.

Crystalline Describing substances that are characterized by a regular, orderly arrangement of atoms.

Curie temperature The temperature above which a material, when heated, ceases to act as a permanent magnet. When cooled below the Curie temperature, a mineral may become a weak permanent magnet with field aligned in the direction of the earth's magnetic field.

Cytoplasm The viscous material within plant and animal cells (other than that within the cell nucleus).

Deep water wave A wave traveling in water sufficiently deep so that there is little or no interaction between the bottom and the orbital motion of the wave. The velocity of a deep-water wave is determined solely by its wavelength.

Density Mass per unit volume of a substance. Usually expressed as grams per cubic centimeter (g/cm^3).

Deposit feeder Animal that obtains food from the sediment on the bottom of the ocean or lake.

Dipolar magnetic field A simple magnetic field such as that produced by a bar magnet. The term *dipolar* indicates that the field resembles one produced by a magnet with *two* poles (one north pole and one south pole).

Dissociation reaction A chemical reaction in which a molecule breaks into two or more smaller fragments (ions). For example, carbonic acid dissociates into hydrogen ion and bicarbonate ion.

$$H_2CO_3 \longrightarrow H^+ + HCO_3^-$$

Dissolved species An ion or molecule dissolved in water.

Divergent boundary A lithospheric plate boundary at which adjacent plates diverge as a result of seafloor spreading. The boundary is marked by an oceanic ridge or rise.

Downwelling Downward motion of surface water into the subsurface, usually as a result of convergence of surface waters in the open ocean as the result of Ekman transport, or impingement of surface currents on the coast.

Dynamic topography A description of the shape of the surface of a large region of the ocean determined from temperature-salinity-depth data at a large number of locations. A contour map of the dynamic topography is useful for estimating the nature of geostrophic currents.

Earthquake focus The point within the earth from which earthquake waves emanate. It is the point at which the break responsible for the earthquake occurs.

Ekman spiral A theoretical consideration of the effect of a steady wind blowing over an ocean of unlimited depth and breadth and of uniform viscosity. The result is a surface flow at 45° to the right of the wind in the Northern Hemisphere. Water at increasing depth will drift in directions increasingly to the right. The net water transport is 90° to the wind and velocity decreases with depth.

Ekman transport The net transport of water set in motion by wind. Owing to the Ekman spiral phenomenon, it is theoretically 90° to the right of the wind in the Northern Hemisphere and 90° to the left of the wind in the Southern Hemisphere.

Electromagnetic radiation Energy that travels as waves or particles with the speed of light. The properties of electromagnetic radiation depend upon its wavelength. Listed in order of decreasing wavelength, electromagnetic radiation includes radio waves, infrared radiation, visible light, ultraviolet radiation, and x-rays.

Elevation The angle between the horizontal and a line from an observer to an object overhead. An object on the horizon has an elevation of 0°. An object directly overhead has an elevation of 90°.

Emergent coast A shoreline developing from the emergence of the ocean floor relative to the ocean surface. It is usually rather straight, and is characterized by marine features such as terraces, usually found at greater depth.

Epicenter The point on the surface of the earth that is directly above the focus of an earthquake.

Epifaunal Pertaining to organisms that live on the ocean bottom, either attached to it or moving freely over it.

Euphotic zone The surface layer of the ocean that receives enough light to support photosynthesis. The bottom of this zone varies, and reaches a maximum value of about 150 m (approx. 500 ft) in the very clearest open ocean water.

Fetch The distance across an open area of water over which the wind blows, generating waves.

Food chain The passage of energy-yielding materials from producers through a sequence of herbivores and carnivores.

Food web A group of interrelated food chains.

Foraminifer A protist (or protozoan) somewhat distantly related to the ameba. Foraminifers secrete microscopic shells of calcite (calcium carbonate) that are the predominant constituents of biogenous sediments in many areas of the ocean.

Foreshore The portion of the shore lying between the normal high-water and low-water marks; the intertidal zone.

Fracture zone An extensive linear zone of unusually irregular ocean floor topography, characterized by large seamounts, steep-sided or asymmetrical ridges, troughs, or long, steep slopes. A fracture zone usually marks an ancient, inactive transform fault zone.

Fully developed sea When wind of a given speed has blown in the same direction over a sufficiently great fetch and for a sufficiently long period of time, increase in either the length of fetch or the time produces no increase in the maximum average wave height. The sea is then called fully developed.

Fungi The kingdom of fungi. In the marine environment they are found living symbiotically with algae as lichen in the intertidal zone and as decomposers of dead organic matter in the open sea. Term used interchangeably with *Mycota*.

Generic name The genus to which an organism belongs.

Geostrophic current A current that develops as the result of the earth's rotation. It is the result of a near balance between gravitational force and the Coriolis effect.

Gigaton (Gton) One billion (10^9) metric tonnes.

Global plate tectonics The processes by which lithospheric plates are created along spreading ridges, and move across the earth's surface to collide with or slide by one another.

Graded bedding Stratification of sediment in which each layer displays a decrease in grain size from bottom to top.

Gram The basic unit of mass of the metric system (approximately 0.035 ounces Avoirdupois)

Great circle The intersection between the surface of the earth and any plane that passes through the center of the earth. The Equator is a great circle, and each meridian is half a great circle. A great circle can be drawn to connect any two points on the surface of the earth, and is the shortest distance between those two points.

Greenhouse effect The heating of the earth's atmosphere that results from the absorption of the earth's outgoing infrared radiation by components of the atmosphere such as CO_2 and H_2O.

Groin A low, artificial structure projecting into the ocean from the shore, built to interfere with longshore transportation of sediment. It usually has the purpose of trapping sand to cause a beach to build up.

Group velocity The speed with which groups of waves move. In deep water, the group velocity is half the wave velocity.

Gulf Stream The high-intensity western boundary current of the North Atlantic Ocean subtropical gyre. It flows in a northerly to northeasterly direction off the east coast of the United States and Canada.

Hachure One of a set of short parallel lines used in map making as an aid in depicting topography and bathymetry. On a bathymetric chart, a closed depression may be indicated by drawing hachures on the deep-water side of the contours surrounding the depression.

Half-spreading rate The rate at which a lithospheric plate moves away from the spreading axis as a result of seafloor spreading. It is equivalent to the rate of creation of a single lithospheric plate.

Hawaiian-Emperor Seamount Chain A dog-leg shaped chain of islands and seamounts extending from Hawaii to a point near the western end of the Aleutian Islands (Alaska) and the southwestern end of the Kamchatka Peninsula (Russia).

Herbivore An animal that eats plants.

Heterotrophic Describing an organism that is not capable of producing its own food by photosynthesis or chemosynthesis. Heterotrophs depend on the organic compounds produced by plants and animals as food.

Histogram A graph of the frequency distribution of observations, usually drawn as a series of bars the lengths of which indicate the proportion of the measurements falling within given ranges.

Homolosine projection A widely used projection for depicting the whole earth. It is a hybrid of two other projections, a sinusoidal projection equatorward of the 40th parallel and a Mollweide projection at higher latitudes. Distortion is minimized by a series of cuts in the projection. These cuts are placed in the oceans to minimize distortion of the continents, or are placed in the continents when the oceans are the main point of interest.

Hot spot A zone of partially molten material within the asthenosphere that moves slowly or not at all relative to the overlying lithosphere. A jet of this molten material rising through the lithosphere can produce a chain of volcanoes such as those of the Hawaiian-Emperor Seamount Chain.

Hydrogen ion The ion that creates the acidic properties of acids. Normally it is written as H^+, but it is in fact associated with varying numbers of water molecules, as H_3O^+, $H_5O_2^+$, etc.

Hydrogenous Referring to those components of ocean sediments that formed by chemical reactions within seawater or by interaction between seawater and rocks or sediments. Examples include manganese nodules and the clay minerals glauconite and (in some occurrences) smectite or montmorillonite.

Hydrographic chart A chart that shows the distribution of water, water depths, and other information useful for navigation.

Hydrography The description of the surface waters (oceans, lakes, or rivers) of the earth, especially for purposes of navigation.

Hypsography The description of the earth's surface in terms of the distribution of elevations with respect to sea level. A cumulative frequency diagram depicting the distribution of elevations of the earth's surface is called a hypsographic curve. Distributions of elevations are also frequently plotted as histograms.

Infaunal Pertaining to animals that live buried in the soft substrate (mud or sand) of the ocean bottom.

Infrared Radiation with wavelengths longer than those of visible light. Warm objects give off energy in the form of infrared radiation. When it is absorbed by other objects it causes them to become warmed.

Initial freezing temperature The temperature at which seawater, upon cooling, just begins to freeze. Unlike fresh water, which freezes at a sharply defined temperature, seawater freezes gradually and progressively as it is cooled.

Ion An atom (or group of atoms) that becomes electrically charged by gaining or losing one or more electrons. Loss of electrons results in a cation (a positively charged ion). Gain of electrons results in an anion (a negatively charged ion).

Isopyncnal A line of constant density on a temperature-salinity diagram.

Isostasy The floating of the lithosphere upon the asthenosphere. At isostatic equilibrium, the downward force of gravity upon the lithosphere is just balanced by the buoyant force produced by the denser asthenosphere upon which the lithosphere floats.

Jetty A structure built from the shore into a body of water to protect a harbor or navigable passage from being shoaled by deposition of longshore drift material.

Juvenile water Water that is coming from the earth's interior to the surface for the first time since the formation of the planet.

Knot A unit of measurement commonly used to describe velocities in the ocean (of ships, current, etc.). One knot equals one nautical mile per hour.

Latitude Location on the earth's surface defined as the angular distance north or south of the Equator. The latitude of the Equator is 0°. That of the North Pole is 90°N and that of the South Pole is 90°S.

Lava — Molten rock that comes to the surface of the earth through an opening in the crust. The term is also used for the volcanic rock that forms when the molten rock solidifies.

Linnaean classification — The taxonomic scheme of classification of biological organisms developed by Carolus Linnaeus. This is the commonly used scheme of *Kingdom, Phylum, . . . Genus, Species.*

Lithogenous — Referring to mineral components of sediment that are derived from erosion of rocks on continents and islands and carried to the ocean by wind or running water. Equivalent to *terrigenous.*

Lithosphere — The outer layer of the earth's structure, including the crust and the upper mantle, to a maximum depth of about 200 km (approx. 124 miles). It is the lithosphere that breaks into the plates that are the major elements of plate tectonics.

Longitude — Location on the earth's surface, defined as angular distance east or west of the Prime Meridian (which passes through Greenwich, England). The Prime Meridian is 0° longitude. 180° is the International Date Line.

Longshore current — A current located in the surf zone and moving parallel to the shore. It is created by breaking waves.

Longshore drift — The movement of sediment along the beach from the breaker line to the top of the swash zone.

Magma — Molten rock within the interior of the earth.

Magnetic anomaly — Distortion of the regular pattern of the earth's magnetic field, resulting from the magnetic properties of the rocks in the vicinity of the measurement. Positive magnetic anomalies refer to magnetic field intensities that are greater than that expected from the earth's magnetic field. Negative magnetic anomalies refer to weaker than expected magnetic field intensities.

Magnetic declination — The angular difference between true north (the direction to the North Pole) and the direction in which a compass needle points. Declination varies from place to place and is expressed in degrees east or west of true north.

Magnetic dipole — A magnet consisting of one north pole and one south pole, such as a simple bar magnet. The earth's magnetic field is approximately like one that would be produced by a magnetic dipole.

Magnetic field intensity — The intensity of the magnetic field measured at any point on the earth. The measured intensity primarily reflects the magnetic field produced by the earth but is also affected by magnetic fields produced by rocks in the vicinity of the measurement.

Magnetic North — The direction to which a magnetic compass needle points.

Magnetic stripes — A pattern of magnetic anomalies commonly observed in the ocean, characterized by alternating bands of positive and negative magnetic anomalies.

Magnetite — An iron oxide mineral with the formula Fe_3O_4. Magnetite is strongly magnetic and, suspended from a string, was the lodestone or compass of early mariners. It is widely present in rocks in small quantities and is responsible for the natural remanent magnetism of many rocks.

Magnetometer — A device that can be towed behind a ship to measure the magnetic field intensity.

Mantle — The zone between the core and the crust of the earth. The mantle is rich in ferromagnesian minerals (those composed chiefly of oxygen, silicon, iron, and magnesium).

Mercator projection A commonly used cylindrical map projection. In a Mercator projection lines of longitude are straight and are perpendicular to lines of latitude. An advantage of the projection is that any course of constant compass direction plots as a straight line. A disadvantage is that areas are greatly distorted at high latitudes.

Meridian Any half great circle terminating at the North Pole and South Pole. A meridian is a line of longitude.

Metaphyta The kingdom of many-celled plants. Term used interchangeably with *Plantae.*

Metazoa The kingdom of many-celled animals. Term used interchangeably with *Animalia.*

Meter The basic unit of length of the metric system (approximately 39.37 inches).

Metric system A system of measurement used in most countries of the world and for many scientific purposes in the United States. The basic unit of length is the meter (approximately 39.37 inches) and the basic unit of mass is the gram (approximately 0.035 ounces).

Monera Kingdom of organisms having nuclear material spread throughout the cell rather than confined within a sheath. Bacteria and blue-green algae.

Monotonic Referring to something that increases or decreases continually and without interruption.

Motile Referring to organisms that are capable of movement.

Mycota The kingdom of fungi. In the marine environment they are found living symbiotically with algae as lichen in the intertidal zone and as decomposers of dead organic matter in the open sea. Term used interchangeably with *Fungi.*

Nautical mile 6076 feet.

Nearshore The part of the coastal zone that extends from the shoreline seaward to the line of breakers.

Negative magnetic anomaly A condition in which the intensity of the magnetic field measured at a location is less than that expected from the normal global dipolar field.

Nektobenthos Those members of the benthos that can actively swim and spend much time off the bottom.

Nekton Pelagic animals, such as adult squid, fish, and mammals, that are active swimmers. They can determine their position in the ocean by swimming.

Neritic environment The region of the oceans overlying the continental shelves.

Neritic sediment Sediment, composed primarily of terrigenous material, deposited upon the continental shelves, slope, and rise.

Nitrogen-fixing bacteria Any of the bacteria that convert atmospheric nitrogen (N_2) into oxides of nitrogen (NO_2, NO_3) usable by algae in their primary production.

Normal magnetic field Describing the magnetic field generated by the earth today and at other times when North and South Magnetic Poles were aligned approximately as they are today, the North Magnetic Pole in the vicinity of the North Rotational Pole and the South Magnetic Pole in the vicinity of the South Rotational Pole.

North Star A star in the Little Dipper that is visible throughout the Northern Hemisphere. It is positioned almost directly above the North Pole, and therefore, unlike other stars, it scarcely appears to move in the sky. It is very useful for navigation. The direction to the North Star is North from any point in the hemisphere. Its elevation is equal to the latitude.

Nutrient Any organic or inorganic compound used by plants in primary production. In the ocean, the most critical nutrients (because of limited supply) are nitrogen and phosphorous compounds, and in some cases, silicon.

Ocean This term is used to describe either the continuous body of water surrounding the continents and covering much of the earth's surface (i.e., the World Ocean) or its major subdivisions (e.g., Atlantic, Pacific, etc.)

Ocean trenches A long, narrow, and deep depression in the floor of the ocean, with relatively steep sides. A trench marks the location where one lithospheric plate is subducted under another.

Oceanic environment The region of the oceans that lie oceanward of the continental shelves.

Oceanic ridges/rises Long, sometimes sinuous, seismically active mountain ranges that extend through all of the major oceans, rising 1 to 3 km above the deep ocean floor. These are the sites of creation of new seafloor in the seafloor spreading process. Where spreading rates are slow, the mountain ranges are relatively steep-sided and are called *oceanic ridges* or *mid-ocean ridges*. Where spreading rates are rapid, the mountain ranges are wider and rise more gradually from the surrounding deep ocean floor and are called *oceanic rises* or *mid-ocean rises.*

Oceanic sediment Sediment deposited on the floor of the deep oceans.

Offshore The comparatively flat portion of the coastal zone extending from the breaker line to the edge of the continental shelf.

Orbital motion The pattern of motion of particles of water as a wave passes.

Organic compound A carbon-containing compound, especially one in which the carbon is in the reduced state or one in which a number of carbon atoms are joined together. The term is used whether the compound was formed in a biological or abiological process.

Organic matter Chemical compounds composed of carbon and hydrogen, and in some cases containing nitrogen, oxygen, sulfur, phosphorous and other constituents. These materials are produced, and in some cases modified, by organisms.

Outgassing The transport of volatiles (e.g., water, atmospheric gases) from the earth's interior to the surface.

Oxidized carbon Carbon in chemical compounds with oxygen and other elements that have a great affinity for electrons. The most common examples of oxidized carbon are CO_2 and bicarbonate ion (HCO_3^-).

Parts per million (ppm) A unit of concentration. Usually expressed in terms of weight unless volume is specified. A solution with a concentration of one ppm contains 1 gm of solute in each million grams of solution. This is equivalent to 1 mg of solute per kg of solution.

Passive margin The margin of a continent that is not significantly deformed by tectonic processes, because it is not a lithospheric plate margin. The Atlantic coast of North America is an example.

Pebble Particles with sizes between 4 mm and 64 mm. Finer than cobbles and coarser than sand.

Pelagic Referring to the open-ocean environment and to organisms living in that environment. The pelagic environment is divided into the neritic province (overlying the continental shelves) and the oceanic province (encompassing the waters overlying the deep sea).

Photosynthesis The process by which green plants convert CO_2, H_2O, and light energy into glucose ($C_6H_{12}O_6$) and O_2.

Phylogenetic Pertaining to the evolutionary pathway of an organism.

Physiographic diagram A map that depicts the shape of the landforms on the earth's surface in a pictorial fashion.

Phytoplankton Plant plankton. These constitute the most important community of primary producers in the ocean.

Plankton Passively drifting or weakly swimming organisms that are dependent on currents to provide motion. Includes primarily microscopic algae, protozoans, and larvae of higher animals.

Plantae The kingdom of many-celled plants. Term used interchangeably with *Metaphyta*.

Polyconic projection A series of conic projections pieced together to create a map with less distortion than a simple conic projection.

Polymodal distribution A frequency distribution of observations that is characterized by three or more ranges of maximum values.

Positive magnetic anomaly A condition in which the intensity of the magnetic field measured at a location is greater than that expected from the normal global dipolar field.

Predator An animal that obtains its food by preying on other living animals.

Primary carnivore An animal that eats herbivores.

Primary producer An organism capable of producing its own food by photosynthesis. An autotroph.

Primary production Synthesis of organic matter from inorganic compounds (primarily CO_2 and H_2O) by the organisms using the process of photosynthesis. Light energy is required for photosynthesis to proceed.

Prime Meridian The meridian of 0° longitude, used as a reference for measuring longitudes. The Prime Meridian passes through the observatory at Greenwich, England, and is also called the Greenwich Meridian.

Projection The depiction of the three-dimensional surface of the earth (the globe) on a plane surface (a map) using a mathematical formula that relates each position on the globe to a position on the plane. Common projections include Mercator projections, conic projections, and planar projections.

Protista Kingdom of organisms that includes all one-celled forms with nuclear material confined within a nuclear sheath. Includes the animal phylum Protozoa and the phyla of algal plants.

Pythagorean Theorem Equation describing the relationship among the sides of a right triangle. $A^2 + B^2 = C^2$ where A and B are the lengths of the sides adjacent to the right angle and C is the length of the hypotenuse (the side opposite the right angle).

Reduced carbon Carbon in most organic (biologically formed) compounds. The formation of glucose from CO_2 in the process of photosynthesis involves the reduction of carbon.

Remanent magnetism The magnetization of a rock that is permanent, in the sense that it is not affected by relatively modest changes in the magnetic field of the environment such as reversal of the earth's magnetic field.

Reservoir A geochemical term used to denote a significant mass of a constituent to which additions may be made and from which withdrawals may occur, either by natural or anthropogenic processes.

Residence time The average length of time an atom of a chemical constituent of seawater remains in the ocean between the time it first enters and the time it is removed. It is a statistical measure, because individual atoms may remain in the ocean for times that are much longer or much shorter than the average.

Respiration The oxidation of organic matter by organisms. This is the process by which the organism obtains the energy stored in the compounds during photosynthesis. Primary reaction products are CO_2 and H_2O.

Reversed magnetic field Describing the magnetic field generated by the earth at times when North and South Magnetic Poles had an orientation approximately the reverse of today's, the North Magnetic Pole in the vicinity of the South Rotational Pole and the South Magnetic Pole in the vicinity of the North Rotational Pole.

Rhumbline A straight line connecting two points on a Mercator projection. Travel from one point to another along a rhumbline is done along a constant compass direction. A rhumbline, however, is the shortest distance between two points only when it is also a great circle. Great circles and rhumblines coincide only along the equator and along meridians.

Salinity A measure of the amount of salt dissolved in seawater. By general agreement among oceanographers it is equal to the total amount of solid material (in grams) dissolved in a kilogram of seawater after certain adjustments are made. Salinity is commonly expressed as parts per thousand (ppt or per mil). Salinity is usually determined by measuring the electrical conductivity of the water, its refractive index, or the concentration of chloride ion (referred to as the chlorinity).

Sand Particles with sizes between 0.0625 mm and 2 mm. Finer than pebbles and coarser than silt.

Sea 1. A subdivision of an ocean, commonly at least partially isolated from the remainder of the ocean by continents or islands (e.g., Mediterranean Sea, Caribbean Sea). 2. A region of the ocean in which the wind generates waves. The waves in this region are a chaotic jumble of many different wavelengths.

Seafloor spreading A process responsible for creation of lithosphere along oceanic ridges. Convective upwelling of magma along the ridges results in the enlargement of the lithospheric plates on either side of the ridge. The plates typically move away from the ridge at velocities of 1 to 10 cm (0.4 to 4 in) per year.

Secondary carnivore An animal that eats primary carnivores. Primary carnivores, in turn, eat herbivores, and herbivores eat plants.

Seiche A standing wave with one node in an enclosed or semi-enclosed body of water. The wave may have a period from a few minutes to a few hours, depending upon the dimensions of the basin.

Sessile Referring to benthic organisms that are permanently attached to the substrate and not free to move about.

Shallow water wave A wave traveling in water sufficiently shallow so that there is very strong interaction between the bottom and the orbital motion of the wave. The velocity of a shallow-water wave is determined solely by the depth of the water.

Shore The region of the coastal zone that extends from the highest level of wave action during storms to the low-water line.

Silt Particles with sizes between 0.004 mm (4 μm) and .0625 mm. Finer than sand and coarser than clay.

Size fraction When the grains making up a sediment are separated according to size, as might be done using sieves, each fraction has an upper and a lower size limit (for example, coarser than 1 mm but less than 2 mm). Fractions of the sediment defined according to their size are called size fractions.

Small circle The intersection between the surface of the earth and any plane that does not pass through the center of the earth. All lines of latitude (other than the Equator) are small circles.

Solute A substance dissolved in a solution. In seawater, the salts are the most important solutes.

Sorting The separation of sedimentary constituents by natural processes according to particle size.

Sounding Measurement of the depth of water at any location. Water depths shown as numbers on charts are often referred to as soundings.

Spectrum The distribution of wavelengths of radiation.

Standing wave A wave, the form of which oscillates vertically without moving forward. The region of maximum vertical motion is the antinode. Between antinodes are nodes, where there is no vertical motion.

Statute mile 5280 feet.

Steady state A condition in which the rate at which a chemical constituent entering the ocean is equal to the rate at which it is removed from the ocean. Thus, while individual atoms of the constituent move into and out of the ocean, the total amount of the constituent does not change with time.

Stony meteorite Extraterrestrial objects composed primarily of silicate minerals. Stony meteorites are the most abundant of meteorites. Other classes of meteorites include the *irons* and the *stony irons.*

Subduction A process by which one lithospheric plate descends beneath another. The surface expression of such a process may be an island arc-trench system or a folded mountain range.

Submarine canyon A steep, V-shaped canyon cut into the continental shelf or slope.

Submergent coast A coast formed by the relative submergence of a landmass. The shoreline is characterized by submerged or partially submerged landforms that developed under subaerial processes. It typically has numerous bays and promontories and is more irregular than an emergent coast.

Subtropical gyre The trade winds and westerly winds initiated in the subtropical regions of all ocean basins, with the influence of the Coriolis effect, set large regions of ocean water in motion. They rotate clockwise in the Northern Hemisphere and counterclockwise in the Southern Hemisphere.

Suspension feeder An animal that obtains food that is suspended in the water around it.

Swash A thin layer of water that washes up over the exposed beach as waves break.

Swell Waves with smooth profiles and long wavelengths. These may travel over long distances with little energy loss.

Terrace A wave-cut bench that has been exposed above sea level as the result of a relative drop in sea level.

Terrestrial biosphere All of the organisms living on the land.

Terrigenous Referring to mineral components of sediment that are derived from erosion of rocks on continents and islands and carried to the ocean by wind or running water. Equivalent to *lithogenous.*

Thermocline A layer in the water in which the temperature decreases sharply with depth. The thermocline lies below the mixed layer of surface and near-surface water and above the zone of cold deep water.

Total spreading rate The rate at which two plates diverge at a spreading center. In a simple case like the Atlantic Ocean, in which subduction does not occur, it is equivalent to the rate at which the size of the ocean increases.

Trace element An element present in seawater in a very low concentration. Some are essential to some organisms.

Transform boundary A boundary between two lithospheric plates marked by shearing motion. That is, the two lithospheric plates do not converge or diverge, but slide past one another.

Transitional wave A wave traveling in water sufficiently shallow so that there is some interaction between the bottom and the orbital motion of the wave. The velocity of a transitional wave is determined both by the depth of the water and the wavelength of the wave.

Trophic level A nourishment level in a food chain. Algal producers constitute the lowest level, followed by herbivores and a series of carnivores at the higher levels.

True North The direction from any point on the surface of the earth to the North Pole.

Turbidite A sediment or rock formed from sediment deposited by a turbidity current. It is typically characterized by both horizontally and vertically graded bedding.

Turbidity current A rapidly moving current consisting of a dense suspension of sediment particles in water. The dense suspension flows along the bottom and can move rapidly down the continental margin under the force of gravity.

Ultraviolet Radiation with wavelengths shorter than those of visible light. Solar energy includes much ultraviolet radiation, most of which is absorbed by the ozone of the atmosphere before it can reach the surface of the earth.

Unimodal distribution A frequency distribution of observations that is characterized by a single range of maximum values.

Upwelling The process by which deep, cold, nutrient-laden water is brought to the surface, usually by the wind divergence of equatorial currents, coastal winds pushing water away from the coast, or the deflection of currents by shallow submarine topographic highs.

Vertical exaggeration Purposeful distortion of vertical dimensions on a map, done to make bathymetry, topography, or other elevational data more readily seen.

Visible light Electromagnetic radiation of wavelengths to which the eye is sensitive. The radiation given off by the sun is centered in the visible part of the electromagnetic spectrum.

Wave dispersion The separation of waves by wavelength as they leave the sea area where they were generated. Waves with longer wavelengths travel faster, and move out ahead of waves with shorter wavelengths.

Wave frequency The number of waves that pass a fixed point in a unit time (usually 1 second). Wave frequency = 1/wave period.

Wave interference The interaction among waves, causing the resulting wave to become either bigger or smaller than the original ones.

Wave period The length of time between the passage of two successive wave crests past a fixed point. Wave period = 1/wave frequency.

Wave tank A device for generating and observing waves in the laboratory.

Wave velocity The speed with which individual waves move.

Wavelength The distance between two corresponding points on successive waves (e.g., from crest to crest).

Well sorted sediment A sediment composed of particles of a narrow size range.

Wind-driven circulation Any movement of ocean water that is driven by winds. That includes most horizontal movements in the surface waters of the world's oceans.

Winnowing The process of removing fine particles from a sediment by blowing or washing them away, leaving a residue of coarser particles behind.

World ocean All of the interconnected parts of the oceans: The Atlantic, Pacific, Indian, and Arctic Oceans, and their associated seas. The term is usually used to stress the interconnections and the arbitrariness of the traditional geographic divisions.

Zooplankton Animal plankton.